γ Gamma

Single and Multiple-Digit Multiplication

Instruction Manual

by Steven P. Demme

1-888-854-MATH (6284) - mathusee.com
sales@mathusee.com

Gamma Instruction Manual: Single and Multiple-Digit Multiplication

Published and distributed by Demme Learning

mathusee.com

1-888-854-6284 or +1 717-283-1448 | demmelearning.com
Lancaster, Pennsylvania USA

ISBN 978-1-60826-081-2
Revision Code 1018

Printed in the United States of America by CJK Group
2 3 4 5 6 7 8 9 10

For information regarding CPSIA on this printed material call: 1-888-854-6284
and provide reference #1118-081519

Building Understanding in Teachers and Students to Nurture a Lifelong Love of Learning

At Math-U-See, our goal is to build understanding for all students.

We believe that education should be relevant, skill-based, and built on previous learning. Because students have a variety of learning styles, we believe education should be multi-sensory. While some memorization is necessary to learn math facts and formulas, students also must be able to apply this knowledge in real-life situations.

Math-U-See is proud to partner with teachers and parents as we use these principles of education to **build lifelong learners.**

Curriculum Sequence

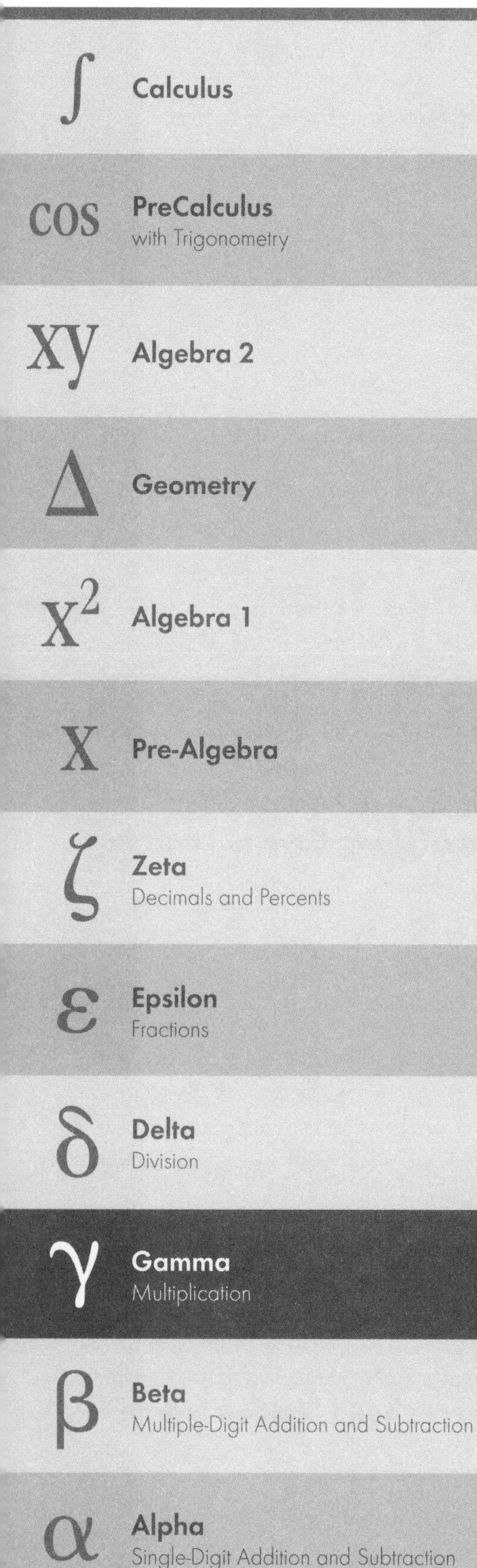

Math-U-See is a complete, K-12 math curriculum that uses manipulatives to illustrate and teach math concepts. We strive toward "Building Understanding" by using a mastery-based approach suitable for all levels and learning preferences. While each book concentrates on a specific theme, other math topics are introduced where appropriate. Subsequent books continuously review and integrate topics and concepts presented in previous levels.

Where to Start

Because Math-U-See is mastery-based, students may start at any level. We use the Greek alphabet to show the sequence of concepts taught rather than the grade level. Go to mathusee.com for more placement help.

Each level builds on previously learned skills to prepare a solid foundation so the student is then ready to apply these concepts to algebra and other upper-level courses.

Major concepts and skills for Gamma:

- Using strategies based on place value and properties of operations to multiply
- Fluently multiplying whole numbers
- Solving for an unknown factor
- Solving abstract and real-world problems involving addition, subtraction, and multiplication
- Measuring and computing area
- Relating concepts of area to multiplication

Additional concepts and skills for Gamma:

- Skip counting as a precursor to multiplication
- Adding and subtracting time in hours and minutes
- Multiplying, adding, and subtracting U.S. currency and standard units of measure
- Representing, recording, and interpreting data
- Understanding basic fractions
- Estimating and solving measurement problems

Find more information and products at mathusee.com

Contents

HOW TO USE

Five Minutes for Success

Welcome to *Gamma.* I believe you will have a positive experience with the unique Math-U-See approach to teaching math. These first few pages explain the essence of this methodology, which has worked for thousands of students and teachers. I hope you will take five minutes and read through these steps carefully.

I am assuming your student has a thorough grasp of addition and subtraction.

If you are using the program properly and still need additional help, you may visit us online at mathusee.com or call us at 888-854-6284.

–**Steve Demme**

The Goal of Math-U-See

The underlying assumption or premise of Math-U-See is that the reason we study math is to apply math in everyday situations. Our goal is to help produce confident problem solvers who enjoy the study of math. These are students who learn their math facts, rules, and formulas and are able to use this knowledge to solve word problems and real-life applications. Therefore, the study of math is much more than simply committing to memory a list of facts. It includes memorization, but it also encompasses learning the underlying concepts of math that are critical to successful problem solving.

Support and Resources

Math-U-See has a number of resources to help you in the educational process.

Many of our customer service representatives have been with us for over 10 years. They are able to answer your questions, help you place your student in the appropriate level, and provide knowledgeable support throughout the school year.

Visit mathusee.com to use our many online resources, find out when we will be in your neighborhood, and connect with us on social media.

More than Memorization

Many people confuse memorization with understanding. Once while I was teaching seven junior high students, I asked how many pieces they would each receive if there were fourteen pieces. The students' response was, "What do we do: add, subtract, multiply, or divide?" Knowing how to divide is important; understanding when to divide is equally important.

The Suggested 4-Step Math-U-See Approach

In order to train students to be confident problem solvers, here are the four steps that I suggest you use to get the most from the Math-U-See curriculum.

Step 1. Prepare for the lesson
Step 2. Present and explore the new concept together
Step 3. Practice for mastery
Step 4. Progress after mastery

Step 1. Prepare for the lesson

Watch the video lesson to learn the new concept and see how to demonstrate this concept with the manipulatives when applicable. Study the written explanations and examples in the instruction manual.

Step 2. Present and explore the new concept together

Present the new concept to your student. Have the student watch the video lesson with you, if you think it would be helpful. The following should happen interactively.

a. **Build:** Use the manipulatives to demonstrate and model problems from the instruction manual. If you need more examples, use the appropriate lesson practice pages.

b. **Write:** Write down the step-by-step solutions as you work through the problems together, using manipulatives.

c. **Say:** Talk through the why of the math concept as you build and write.

Give as many opportunities for the student to "Build, Write, Say" as necessary until the student fully understands the new concept and can demonstrate it to you confidently. One of the joys of teaching is hearing a student say, *"Now I get it!"* or *"Now I see it!"*

Step 3. Practice for mastery

Using the lesson practice problems from the student workbook, have students practice the new concept until they understand it. It is one thing for students to watch someone else do a problem; it is quite another to do the same problem themselves. Together complete as many of the lesson practice pages as necessary (not all pages may be needed) until the student understands the new concept, demonstrating confident mastery of the skill. Remember, to demonstrate mastery, your student should be able to teach the concept back to you using the Build, Write, Say method. Give special attention to the word problems, which are designed to apply the concept being taught in the lesson. If your student needs more assistance, go to mathusee.com to find review tools and other resources.

Step 4. Progress after mastery

Once mastery of the new concept is demonstrated, advance to the systematic review pages for that lesson. These worksheets review the new material as well as provide practice of the math concepts previously studied. If the student struggles, reteach these concepts to maintain mastery. If students quickly demonstrate mastery, they may not need to complete all of the systematic review pages.

In the 2012 student workbook, the last systematic review page for each lesson is followed by a page called "Application and Enrichment." These pages provide a way for students to review and use their math skills in a variety of different formats. You may decide how useful these activity pages are for your particular student.

Now you are ready for the lesson tests. These were designed to be an assessment tool to help determine mastery, but they may also be used as extra worksheets. Your student will be ready for the next lesson only after demonstrating mastery of the new concept and maintaining mastery of concepts found in the systematic review worksheets.

Tell me, I forget. Show me, I understand. Let me do it, I remember.
–Ancient Proverb

To this Math-U-See adds, *"Let me teach it, and I will have achieved mastery!"*

Length of a Lesson

How long should a lesson take? This will vary from student to student and from topic to topic. You may spend a day on a new topic, or you may spend several days. There are so many factors that influence this process that it is impossible to predict the length of time from one lesson to another. I have spent three days on a lesson, and I have also invested three weeks in a lesson. This experience occurred in the same book with the same student. If you move from lesson to lesson too quickly without the student demonstrating mastery, the student will become overwhelmed and discouraged as he or she is exposed to more new material without having learned previous topics. If you move too slowly, the student may become bored and lose interest in math. I believe that as you regularly spend time working along with the student, you will sense the right time to take the lesson test and progress through the book.

By following the four steps outlined above, you will have a much greater opportunity to succeed. Math must be taught sequentially, as it builds line upon line and precept upon precept on previously-learned material. I hope you will try this methodology and move at the student's pace. As you do, I think you will be helping to create a confident problem solver who enjoys the study of math.

LESSON 1

Rectangles, Factors, and Products

The word ***rectangle*** means "right angle." It comes from the Latin word "rectus," which means "right." German, French, Spanish, and other languages have similar words with the same meaning. A ***right angle*** is a square corner. A closed shape with four square corners is a rectangle. Look around for examples of rectangles.

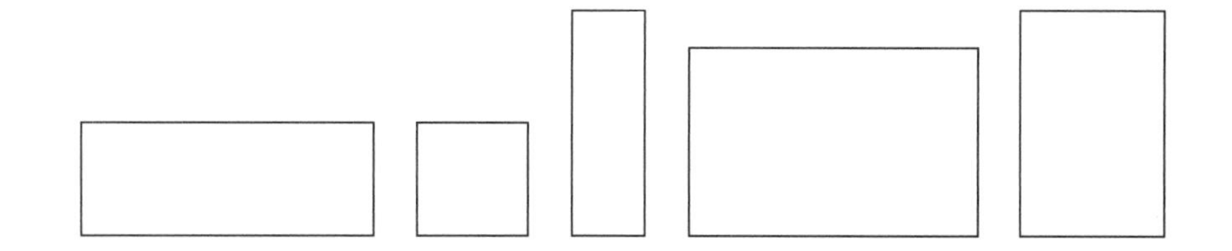

A ***square*** is a special kind of rectangle. It has four right angles, so it is a rectangle, but it also has four sides that are the same length, so it is also a square.

A rectangle is measured by its dimensions. A ***dimension*** is the length of a side. In the following picture, the over dimension is three, and the up dimension is two. The dimensions tell how long the sides are.

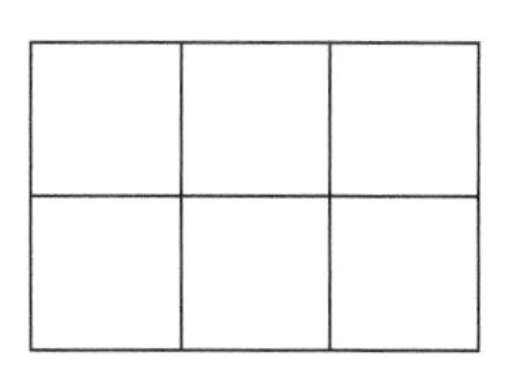

The rectangle also has area. A square that is one unit long on each side is a ***unit square***. We say it has one square unit of area and use it to measure the area of larger shapes. The area tells how many unit squares are needed to fill the rectangle. In this rectangle, the area is six square units.

What is the over dimension and the up dimension? What is the area?

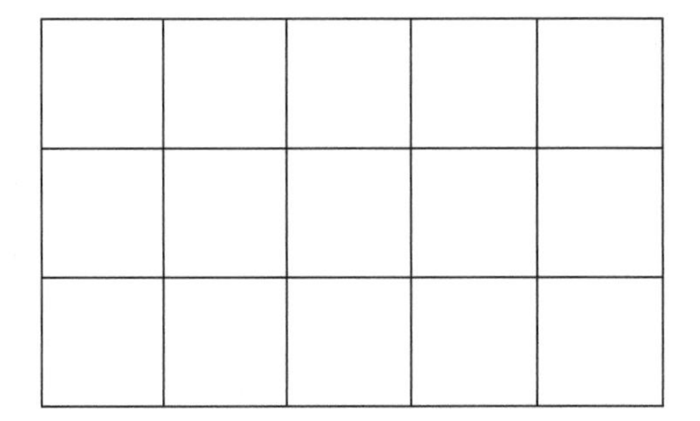

The over dimension is 5, and the up dimension is 3. The area is 15 square units.

Later we will use the words *factor* instead of dimension and *product* instead of area. By learning the skip-counting facts and how to build rectangles, we are laying a foundation for multiplication.

On the worksheets there are rectangles that look like the figures below. Write the dimensions of each rectangle in the parentheses and the area in the oval.

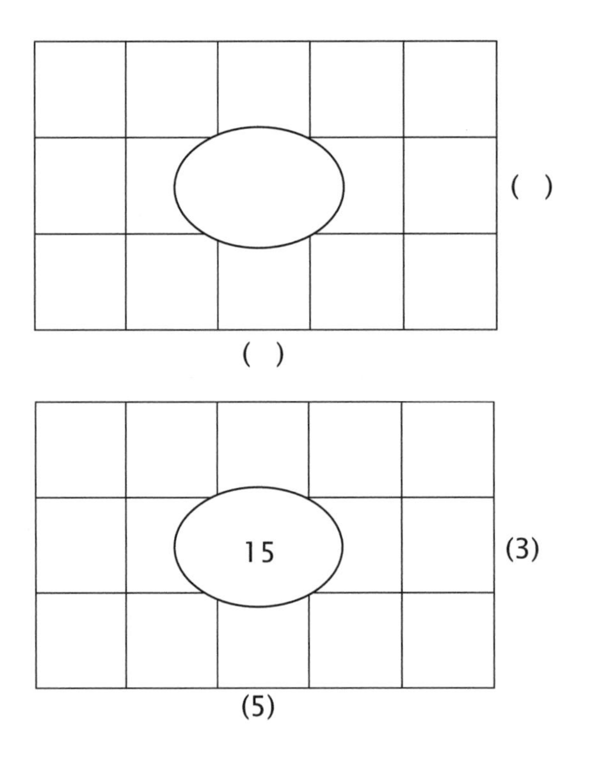

Example 1

Build a rectangle with dimensions over 10 and up 5. We can write this as (10)(5). Now find the area.

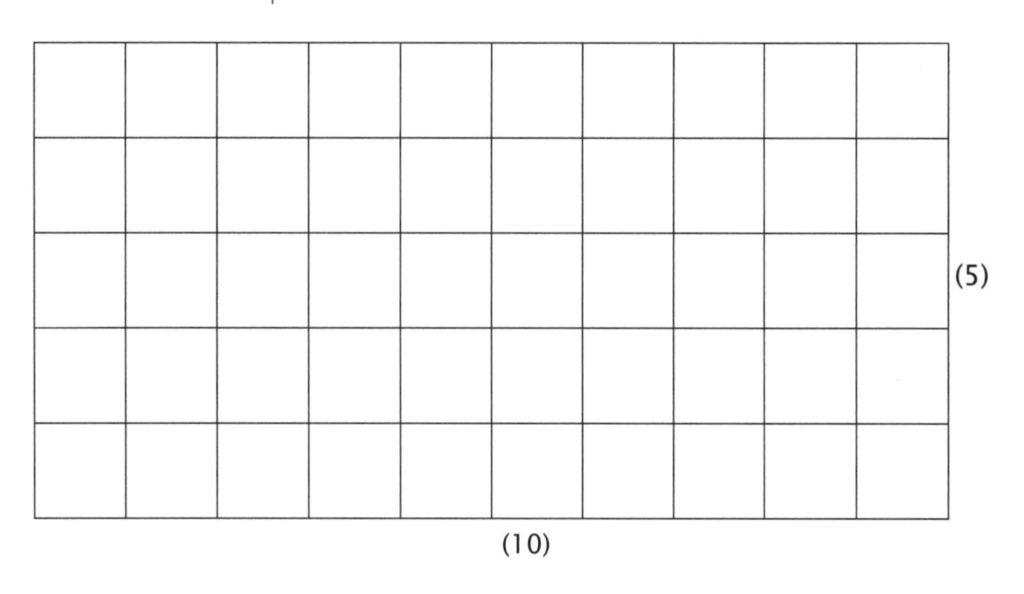

Skip counting by 5 (5-10-15-20-25-30) or by 10 (10-20-30-40-50), we find an area of 50.

Another way to show this is to put a piece of paper over the rectangle in order to emphasize that the dimensions are 10 and 5. In other words, the dimensions are the lengths of the sides. Lift off the paper, and you can see that the area is the same as the product.

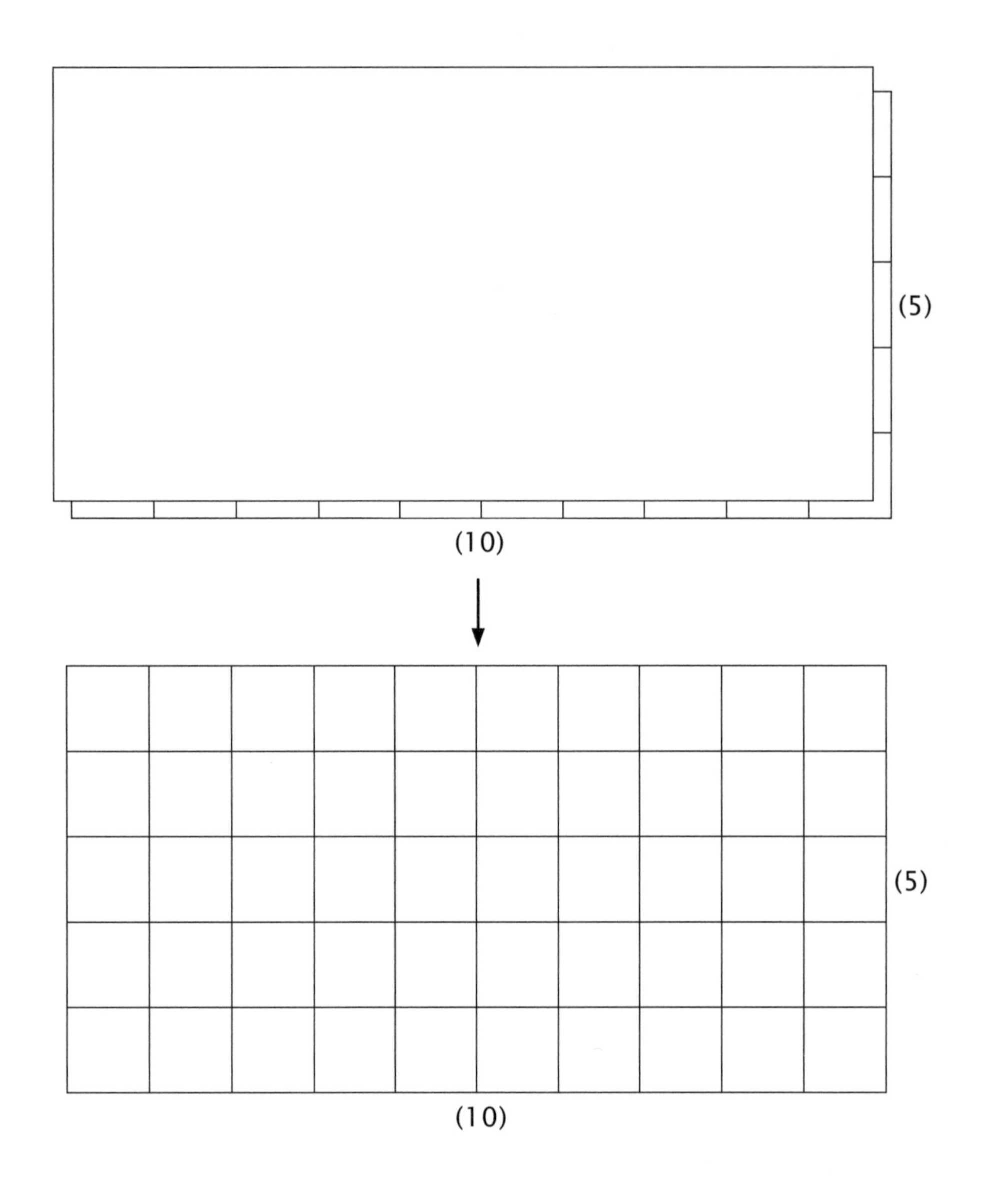
(5)
(10)
(5)
(10)

LESSON 2

Multiply by 1 and 0, Commutative Property

Multiplication is fast adding of the same number. Tell the student that one multiplied by three is simply one counted three times, or 1 + 1 + 1. The facts in this lesson are familiar, but the symbolism is new. We use three ways to illustrate multiplication in this book: "×" (the most common), the floating dot " ·," and two parentheses side by side ( )( ). Although it is not introduced in *Gamma*, the symbol "*" is commonly used for multiplication by calculators and online programs. So far, the student has learned only one symbol for addition and one for subtraction, so comment that there are several symbols for multiplication.

You may verbalize multiplication several ways. The problem 1 × 3 could be "one counted three times," "one times three," or "one multiplied by three." Since multiplication is fast adding of the same number, the problem can also be shown as 1 + 1 + 1 = 3. In Example 1, 1 × 3 is illustrated with a rectangle, and the three ways of writing it are given.

Example 1

$$1 \times 3 = 3$$
$$1 \cdot 3 = 3$$
$$(1)(3) = 3$$

A student may ask whether 1 × 3 is the same as 3 × 1. This is a golden opportunity to teach that multiplication is ***commutative***; that is, you can change the order of the factors without changing the product. To show this, simply build the rectangle vertically and then horizontally to show that both are the same rectangle, except that one is "standing up" and one is "lying down."

Just as in addition and subtraction, the problems themselves may be written either vertically, as one times three, or horizontally, as three times one. When a multiplication problem is written vertically, the number on top is the number being counted. You may see it referred to as the ***multiplicand***. The number that tells how many times we are counting is referred to as the ***multiplier***. Because the answer to a multiplication problem is the same no matter which order we use, it is more useful at this level to use the word ***factor*** to refer to both the multiplier and the multiplicand.

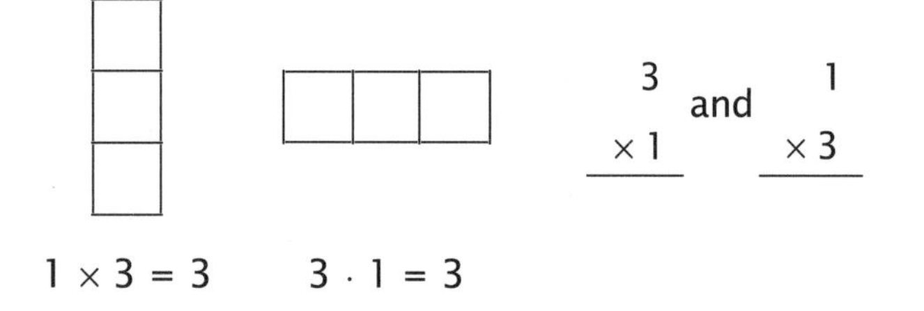

Multiplication by zero is easier than you may think. When you verbalize it, mention the zero factor first, as this makes it easier to grasp. The problem 0×5 is zero counted five times, or $0 + 0 + 0 + 0 + 0$, which is obviously 0. I like to use a word problem to make this clear. I say, "Suppose you didn't earn any money today or for the next four days; how much would you have?" "You would have $0 + 0 + 0 + 0 + 0$, or zero counted five days, or zero money." Use similar stories to illustrate this concept. Demonstrating multiplication by zero with the blocks is somewhat abstract. However, you can point out that if five is counted zero times, there are no five blocks.

This is a good time to show all of the multiplication facts on a chart. After the student learns several facts, color or circle those facts to encourage him in his progress. We put a small chart in each of the lessons so that you can circle the facts currently being studied, as well as those already learned. There is a chart for the student after the lesson 2 worksheets in the student workbook.

0 × 0	0 × 1	0 × 2	0 × 3	0 × 4	0 × 5	0 × 6	0 × 7	0 × 8	0 × 9	0 × 10
1 × 0	1 × 1	1 × 2	1 × 3	1 × 4	1 × 5	1 × 6	1 × 7	1 × 8	1 × 9	1 × 10
2 × 0	2 × 1	2 × 2	2 × 3	2 × 4	2 × 5	2 × 6	2 × 7	2 × 8	2 × 9	2 × 10
3 × 0	3 × 1	3 × 2	3 × 3	3 × 4	3 × 5	3 × 6	3 × 7	3 × 8	3 × 9	3 × 10
4 × 0	4 × 1	4 × 2	4 × 3	4 × 4	4 × 5	4 × 6	4 × 7	4 × 8	4 × 9	4 × 10
5 × 0	5 × 1	5 × 2	5 × 3	5 × 4	5 × 5	5 × 6	5 × 7	5 × 8	5 × 9	5 × 10
6 × 0	6 × 1	6 × 2	6 × 3	6 × 4	6 × 5	6 × 6	6 × 7	6 × 8	6 × 9	6 × 10
7 × 0	7 × 1	7 × 2	7 × 3	7 × 4	7 × 5	7 × 6	7 × 7	7 × 8	7 × 9	7 × 10
8 × 0	8 × 1	8 × 2	8 × 3	8 × 4	8 × 5	8 × 6	8 × 7	8 × 8	8 × 9	8 × 10
9 × 0	9 × 1	9 × 2	9 × 3	9 × 4	9 × 5	9 × 6	9 × 7	9 × 8	9 × 9	9 × 10
10 × 0	10 × 1	10 × 2	10 × 3	10 × 4	10 × 5	10 × 6	10 × 7	10 × 8	10 × 9	10 × 10

Word Problem Tips

Parents often find it challenging to teach children how to solve word problems. Here are some suggestions for helping your student learn this important skill.

The first step is to realize that word problems require both reading and math comprehension. Don't expect a child to be able to solve a word problem if he does not thoroughly understand the math concepts involved. On the other hand, a student may have a math skill level that is stronger than his or her reading comprehension skills. Below are a number of strategies to improve comprehension skills in the context of story problems. You may decide which ones work best for you and your child.

Strategies for word problems:

1. Ignore numbers at first and read the story. It may help some students to read the question aloud. Every word problem tells a story. Before deciding what math operation is required, let the student retell the story in his own words. Who is involved? Are they receiving gifts, losing something, or dividing a treat?

2. Relate the story to real life, perhaps by using names of family members or friends. For some students, this makes the problem more interesting and relevant.

3. Build, draw, or act out the story. Use the blocks or actual objects when practical. Especially in the lower levels, you may require the student to use the blocks for word problems, even when the facts have been learned. Don't be afraid to use a little drama as well. The purpose is to make it as real and meaningful as possible.

4. Look for the common language used in a particular kind of problem. Pay close attention to the word problems on the lesson practice pages, as they model different kinds of language that may be used for the new concept just studied. For example, "altogether" often indicates addition. These "key words" can be useful clues, but they should not be a substitute for understanding.

5. Look for practical applications that use the concept and ask questions in that context.

6. Have the student invent word problems to illustrate the number problems from the lesson.

Cautions:

1. Unneeded information may be included in the problem. For example, we may be told that Suzie is eight years old, but the eight is irrelevant when adding up the number of gifts she received.

2. Some problems may require more than one step to solve. Model these questions carefully.

3. There may be more than one way to solve some problems. Experience will help the student choose the easier or preferred method.

4. Estimation is a valuable tool for checking an answer. If an answer is unreasonable, it is possible that the wrong method was used to solve the problem.

Multiplication Facts Sheet

0 × 0	0 × 1	0 × 2	0 × 3	0 × 4	0 × 5	0 × 6	0 × 7	0 × 8	0 × 9	0 × 10
1 × 0	1 × 1	1 × 2	1 × 3	1 × 4	1 × 5	1 × 6	1 × 7	1 × 8	1 × 9	1 × 10
2 × 0	2 × 1	2 × 2	2 × 3	2 × 4	2 × 5	2 × 6	2 × 7	2 × 8	2 × 9	2 × 10
3 × 0	3 × 1	3 × 2	3 × 3	3 × 4	3 × 5	3 × 6	3 × 7	3 × 8	3 × 9	3 × 10
4 × 0	4 × 1	4 × 2	4 × 3	4 × 4	4 × 5	4 × 6	4 × 7	4 × 8	4 × 9	4 × 10
5 × 0	5 × 1	5 × 2	5 × 3	5 × 4	5 × 5	5 × 6	5 × 7	5 × 8	5 × 9	5 × 10
6 × 0	6 × 1	6 × 2	6 × 3	6 × 4	6 × 5	6 × 6	6 × 7	6 × 8	6 × 9	6 × 10
7 × 0	7 × 1	7 × 2	7 × 3	7 × 4	7 × 5	7 × 6	7 × 7	7 × 8	7 × 9	7 × 10
8 × 0	8 × 1	8 × 2	8 × 3	8 × 4	8 × 5	8 × 6	8 × 7	8 × 8	8 × 9	8 × 10
9 × 0	9 × 1	9 × 2	9 × 3	9 × 4	9 × 5	9 × 6	9 × 7	9 × 8	9 × 9	9 × 10
10 × 0	10 × 1	10 × 2	10 × 3	10 × 4	10 × 5	10 × 6	10 × 7	10 × 8	10 × 9	10 × 10

LESSON 3

Skip Count by 2, 5, and 10

Skip counting is counting groups of the same number quickly. For example, to skip count by three, you would skip the one and the two and say "three," skip the four and the five and say "six," and follow with "9–12–15–18," etc. Skip counting by seven is 7–14–21–28–35–42–49–56–63–70.

Here are five reasons for learning skip counting:

1. Skip counting lays a solid foundation for learning the multiplication facts. The problem 3 + 3 + 3 + 3 can be written as 3 × 4. If you can skip count, you can say "3–6–9–12." Then you could read 3 × 4 as "three counted four times is twelve." As you learn your skip-counting facts, you are learning all of the products of the multiplication facts in order. Multiplication is fast adding of the same number, and skip counting illustrates this beautifully. You can think of multiplication as a shortcut to the skip-counting process. Consider 3 × 5. I could skip count by three five times (3–6–9–12–15) to come up with the solution. Alternately, after I learn my facts, I can say, "Three counted five times is fifteen." The latter is much faster.

2. Skip counting teaches the concept of multiplication. I had a teacher tell me that her students had successfully memorized their facts but didn't understand the concept. After she had taught them skip counting, they comprehended what they had learned. I used to say that multiplying was fast adding. In reality, it is fast adding of the same number. I can't multiply to find the solution to 1 + 4 + 6 + 9, but I can multiply to

solve 4 + 4 + 4. Skip counting reinforces and teaches the concept of multiplication.

3. As a skill in itself, multiple counting is helpful. A pharmacist attending a workshop told me he skip counted when counting pills as they went into the bottles. Another man said he used the same skill for counting inventory at the end of every workday.

4. It teaches you the multiples of a number, which are so important when making equivalent fractions and finding common denominators. For example, 2/5 = 4/10 = 6/15 = 8/20. The numbers 2–4–6–8 are multiples of 2, and 5–10–15–20 are multiples of 5.

5. In the student book, skip counting is reviewed in sequence by asking students to fill in the blanks: __, __, __, 12, 15, __, __, __, 27, 30. This encourages them to find patterns in math, and patterns are key to understanding this logical and important subject.

One way to learn the skip counting facts is with the *Skip Count and Addition Songbook.* Included is a CD with the skip count songs from the twos to the nines.

Another way to teach skip counting is by counting regularly and then beginning to skip some of the numbers. Build a rectangle with two ten bars or use the drawings in the student workbook. Begin by counting each square: 1–2–3–4–5–6 . . . through 20. After this sequence is learned, skip the first number and just count the second: 2–4–6 . . . through 20. This is skip counting. You know it as counting by two.

When first introducing this, you might try pointing to each square and, as you count the first number quietly, ask the student(s) to say the second number loudly. Continue this practice, saying the number more quietly each time until you are just pointing to the first block silently while the student says the number loudly when you point to the second square. The student(s) see it, hear it, say it, and should write the facts 2–4–6 . . . 20 as well.

Example 1

Skip count and write the numbers on the lines. Say each number out loud as you count and write. After filling in the blanks, write the numbers in the spaces provided beneath the figure. The solution follows at the bottom of the page.

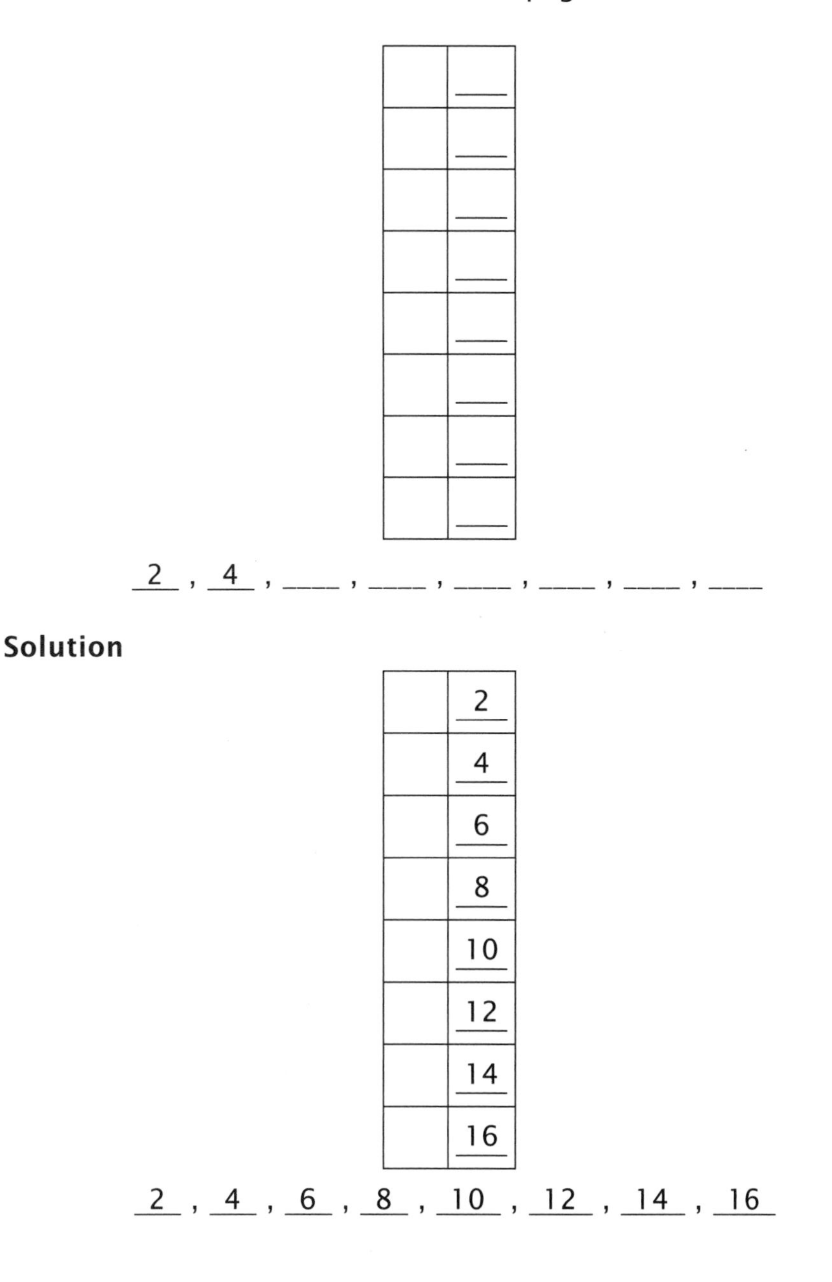

Skip Count by 5

We'll also review skip counting by five in this lesson. Use the same techniques to introduce and teach this important skill as you did for the twos. Some practical examples are fingers on one hand, toes on a foot, cents in a nickel, players on a basketball team, and sides of a pentagon. We can also remind the student that one nickel has the same value as five pennies and apply the skill of counting by five to find out how many cents are in several nickels.

Example 2

As with the twos, skip count and write the numbers on the lines. Say each number out loud as you count and write. After filling out the rectangle, write the numbers in the spaces provided beneath the figure.

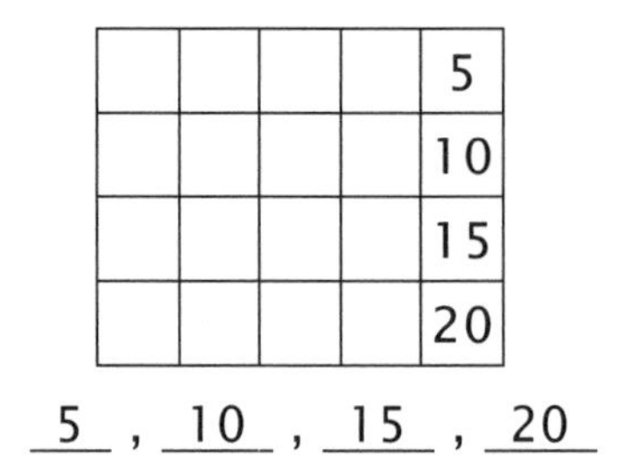

5 , 10 , 15 , 20

Notice that skip counting by five is a way of illustrating multiplication. The rectangle above shows five counted four times or 5×4, which is 20. When students learn all of these skip counting facts, they have learned all of the answers to the five facts. Please don't proceed to the next lesson until the student can skip count by five to 50.

Example 3

Fill in the missing information on the lines.

5 , ___ , 15 , ___ , ___ , 30 , ___ , 40 , 45 , ___

Solution

5 , 10 , 15 , 20 , 25 , 30 , 35 , 40 , 45 , 50

Skip Count by 10

After you learn the twos and fives, review skip counting by ten. Some practical examples are fingers on both hands, toes on both feet, and cents in a dime.

LESSON 4

Multiply by 2, 1 Quart = 2 Pints

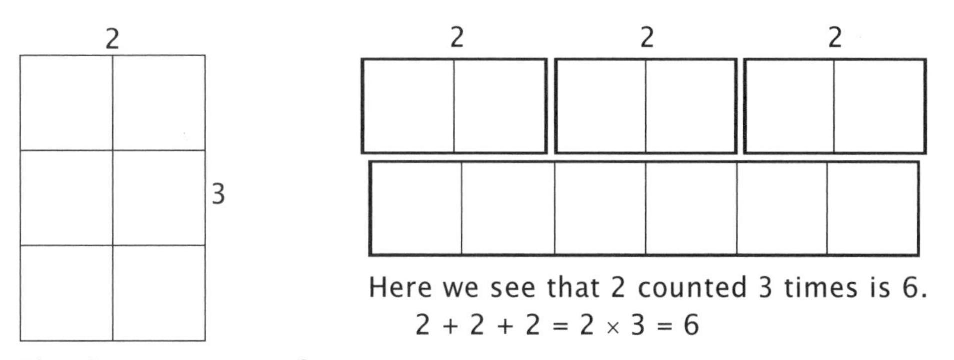

The dimensions, or factors, are 2 and 3, and the area, or product, is 6.

A rectangle has ***dimensions*** and ***area***. The dimensions of this rectangle are two and three. Some say the base is two and the height is three, and others say the length is three and the width is two. Both ways are correct. In this course, we will say the "over" dimension is two, and the "up" dimension is three.

Besides dimensions, a rectangle also has area. In this rectangle, the area would be six square units. Instead of using the term dimensions, we are going to call the over and the up ***factors***. Instead of area, we will use the word ***product***. The problem $2 \times 3 = 6$ is illustrated with the rectangle above. This may be verbalized as "two counted three times," "two times three," or "two multiplied by three."

Remember that, because of the Commutative Property of Multiplication, we can also describe the rectangle as $3 \times 2 = 6$.

Now that the student can skip count by two, we begin memorizing the specific two facts. One way to show them is a progression, with the blocks increasing by two each time. If you use the colored unit bars to represent 2, 4, 6, and 8, a pattern emerges that is pictured below. Notice this as you move from right to left: orange (2)–yellow (4)–violet (6)–brown (8). The pattern repeats beginning with 12.

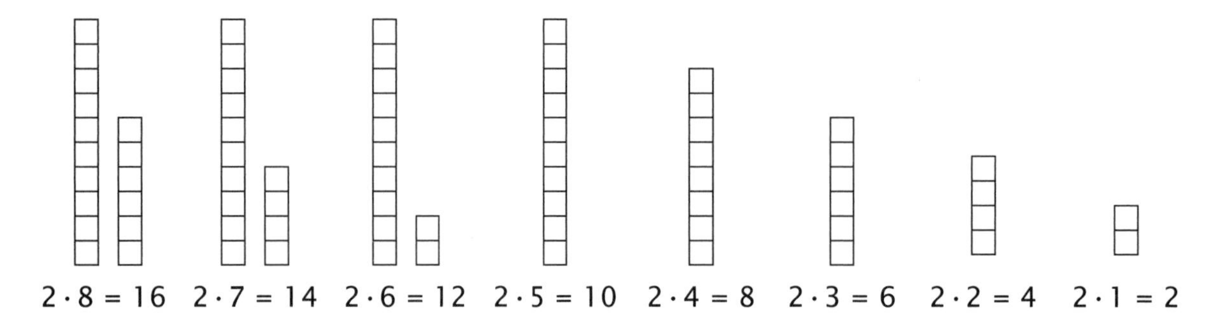

Another way to show the two facts is on a number chart. Circling all of the products for the two facts, or multiples of two, reveals an interesting pattern that corresponds with the blocks above.

(0) 1 (2) 3 (4) 5 (6) 7 (8) 9
(10) 11 (12) 13 (14) 15 (16) 17 (18) 19
(20)

We may also think of a multiplication problem as a size comparison of blocks. One of the numbers that we say when we skip count by two is 12. Writing this as a multiplication problem gives us $2 \times 6 = 12$. An amount that is two times as much as six is 12. Turning the equation around gives us $12 = 2 \times 6$. Make sure that the student understands that both equations are true and mean the same thing. We can verbalize $12 = 2 \times 6$ as "Twelve is two times greater than six." This is very clear when we use the blocks. Build a rectangle that shows 2×6 and take it apart to show that it is made up of two sixes. Be sure the student understands that "two times greater than six" is multiplication, but "two greater than six" is addition.

When the student has learned the skip counting facts and is able to say them independently, he or she knows all of the multiples or products of two. This is a good time to put the factors with the products. Say the factors slowly and then ask the student to say the products by skip counting. For example, you say "two counted one time" or "two times one," and the student says "two." You continue by saying "two times two," and the student says "four."

Here are the two facts with the corresponding products.

0	2	4	6	8	10	12	14	16	18	20
2 × 0	2 × 1	2 × 2	2 × 3	2 × 4	2 × 5	2 × 6	2 × 7	2 × 8	2 × 9	2 × 10
		↑			↑				↑	
		2 counted 2 times			2 counted 5 times				2 counted 9 times	

0 × 0	0 × 1	0 × 2	0 × 3	0 × 4	0 × 5	0 × 6	0 × 7	0 × 8	0 × 9	0 × 10
1 × 0	1 × 1	1 × 2	1 × 3	1 × 4	1 × 5	1 × 6	1 × 7	1 × 8	1 × 9	1 × 10
2 × 0	2 × 1	2 × 2	2 × 3	2 × 4	2 × 5	2 × 6	2 × 7	2 × 8	2 × 9	2 × 10
3 × 0	3 × 1	3 × 2	3 × 3	3 × 4	3 × 5	3 × 6	3 × 7	3 × 8	3 × 9	3 × 10
4 × 0	4 × 1	4 × 2	4 × 3	4 × 4	4 × 5	4 × 6	4 × 7	4 × 8	4 × 9	4 × 10
5 × 0	5 × 1	5 × 2	5 × 3	5 × 4	5 × 5	5 × 6	5 × 7	5 × 8	5 × 9	5 × 10
6 × 0	6 × 1	6 × 2	6 × 3	6 × 4	6 × 5	6 × 6	6 × 7	6 × 8	6 × 9	6 × 10
7 × 0	7 × 1	7 × 2	7 × 3	7 × 4	7 × 5	7 × 6	7 × 7	7 × 8	7 × 9	7 × 10
8 × 0	8 × 1	8 × 2	8 × 3	8 × 4	8 × 5	8 × 6	8 × 7	8 × 8	8 × 9	8 × 10
9 × 0	9 × 1	9 × 2	9 × 3	9 × 4	9 × 5	9 × 6	9 × 7	9 × 8	9 × 9	9 × 10
10 × 0	10 × 1	10 × 2	10 × 3	10 × 4	10 × 5	10 × 6	10 × 7	10 × 8	10 × 9	10 × 10

1 Quart = 2 Pints

Measurement is a good place to apply multiplying by two. Begin with two one-pint containers and one one-quart container. Fill up the one-quart container and empty it into the two one-pint containers. The student can see that one quart equals two pints. You can also show the converse by emptying the two pints into the one quart. Use multiplying by two to find the number of pints in several quarts.

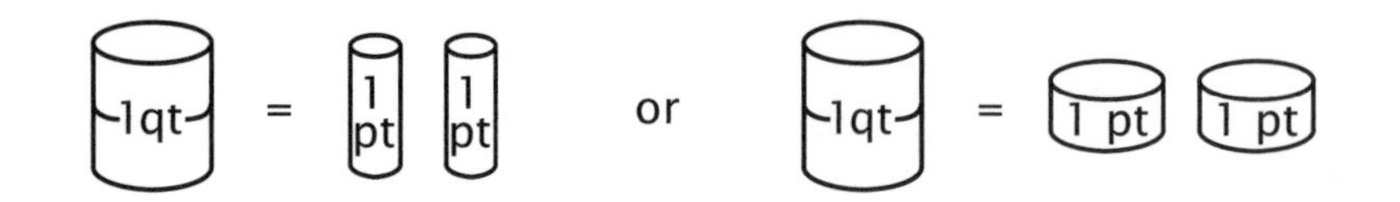

Example 1
How many pints are in three quarts?

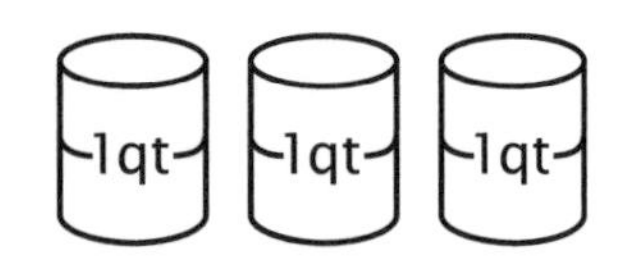

3 quarts × 2 (pints in one quart) = 6 pints
There are six pints in three quarts.

Example 2
How many pints are in the quarts that are shown?

LESSON 5

Multiply by 10, 10¢ = 1 Dime

When multiplying by 10, encourage the student to look for patterns. Notice that whenever you multiply 10 times any number, the answer is that number plus a zero. That is because 10 is made up of a "1" digit which means one ten and a "0" digit that means zero units. Thus, 4×10 is the same as 4×1 ten = 4 tens *and* 4×0 units = 0 units, or 40. The "ty" in forty stands for 10.

To make sure the student has this concept, I like to ask, "What is banana times 10?" The answer is "banana tens," written "banana 0" and pronounced "banana-ty." These are easy facts to learn and remember, but don't take them for granted. Make sure they are mastered using any of the techniques shown below.

On the worksheets, there have been rectangles where the student wrote in the fact at the end of the line in the space with an underline. These can be put to another use by adding the multiplication problem that corresponds to the multiple of 10. Here are a few examples.

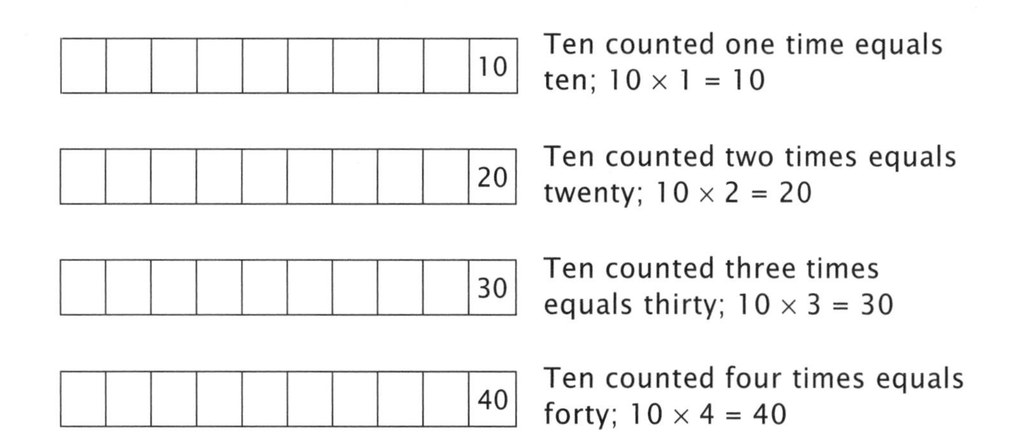

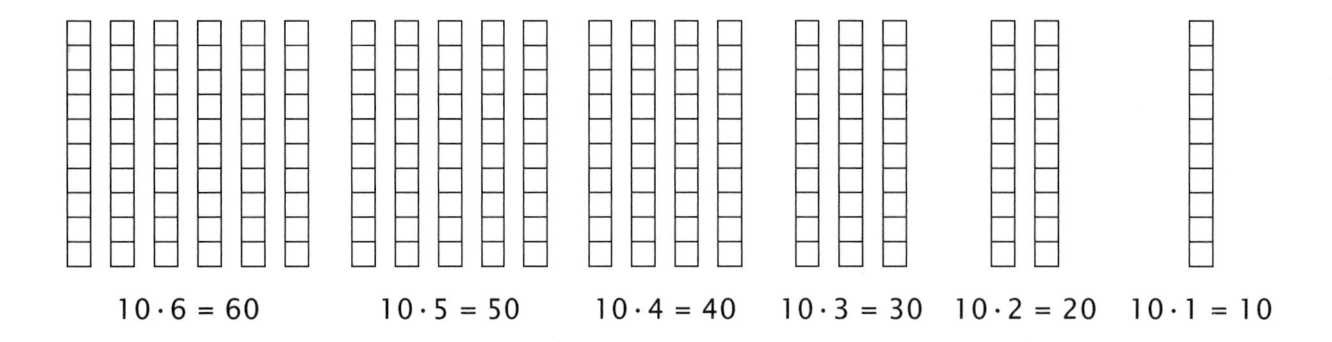

Another way to show this is on a number chart. Circling all of the 10 facts, or multiples of 10, reveals the pattern that corresponds to the blocks above.

(0)	1	2	3	4	5	6	7	8	9
(10)	11	12	13	14	15	16	17	18	19
(20)	21	22	23	24	25	26	27	28	29
(30)	31	32	33	34	35	36	37	38	39
(40)	41	42	43	44	45	46	47	48	49
(50)	51	52	53	54	55	56	57	58	59
(60)	61	62	63	64	65	66	67	68	69
(70)	71	72	73	74	75	76	77	78	79
(80)	81	82	83	84	85	86	87	88	89
(90)	91	92	93	94	95	96	97	98	99
(100)									

Each ten fact can be built in the shape of a rectangle. Whenever illustrating with the blocks, also write the problem and say it as you build.

10 counted 5 times is the same as 50; 10 times 5 equals 50; 10 over and 5 up is 50.

We may also think of a multiplication problem as a size comparison of the blocks. One of the numbers that we say when we skip count by 10 is 30. Writing this as a multiplication problem gives us $10 \times 3 = 30$. Turning the equation around gives us $30 = 10 \times 3$. We can verbalize $30 = 10 \times 3$ as "Thirty is ten times greater than three." Use the blocks to illustrate this if you wish.

Counting by 10 is the first step. After this is accomplished, say the factors slowly and ask the student to say the product. For example, you say "ten counted one time," or "ten times one," and the student says "ten." Continue by saying "ten times two" and having the student say "twenty." (I often have the student say "two-ty" as well as twenty to show the meaning behind the words.) Proceed through all the facts sequentially, just as when the student learned to count by 10.

Here are the 10 facts with the corresponding products.

$\frac{0}{(10)(0)}$ $\frac{10}{(10)(1)}$ $\frac{20}{(10)(2)}$ $\frac{30}{(10)(3)}$ $\frac{40}{(10)(4)}$ $\frac{50}{(10)(5)}$ $\frac{60}{(10)(6)}$ $\frac{70}{(10)(7)}$ $\frac{80}{(10)(8)}$ $\frac{90}{(10)(9)}$ $\frac{100}{(10)(10)}$

↑ 10 counted 1 time

↑ 10 counted 4 times

↑ 10 counted 9 times

0 × 0	0 × 1	0 × 2	0 × 3	0 × 4	0 × 5	0 × 6	0 × 7	0 × 8	0 × 9	0 × 10
1 × 0	1 × 1	1 × 2	1 × 3	1 × 4	1 × 5	1 × 6	1 × 7	1 × 8	1 × 9	1 × 10
2 × 0	2 × 1	2 × 2	2 × 3	2 × 4	2 × 5	2 × 6	2 × 7	2 × 8	2 × 9	2 × 10
3 × 0	3 × 1	3 × 2	3 × 3	3 × 4	3 × 5	3 × 6	3 × 7	3 × 8	3 × 9	3 × 10
4 × 0	4 × 1	4 × 2	4 × 3	4 × 4	4 × 5	4 × 6	4 × 7	4 × 8	4 × 9	4 × 10
5 × 0	5 × 1	5 × 2	5 × 3	5 × 4	5 × 5	5 × 6	5 × 7	5 × 8	5 × 9	5 × 10
6 × 0	6 × 1	6 × 2	6 × 3	6 × 4	6 × 5	6 × 6	6 × 7	6 × 8	6 × 9	6 × 10
7 × 0	7 × 1	7 × 2	7 × 3	7 × 4	7 × 5	7 × 6	7 × 7	7 × 8	7 × 9	7 × 10
8 × 0	8 × 1	8 × 2	8 × 3	8 × 4	8 × 5	8 × 6	8 × 7	8 × 8	8 × 9	8 × 10
9 × 0	9 × 1	9 × 2	9 × 3	9 × 4	9 × 5	9 × 6	9 × 7	9 × 8	9 × 9	9 × 10
10 × 0	10 × 1	10 × 2	10 × 3	10 × 4	10 × 5	10 × 6	10 × 7	10 × 8	10 × 9	10 × 10

10¢ = 1 Dime

A good place to apply multiplication by ten is with money. We've learned that 10¢ is the same as one dime, so we can ask how many pennies are the same as six dimes or how many cents are in six dimes. The answer is 6 × 10¢, or 60¢.

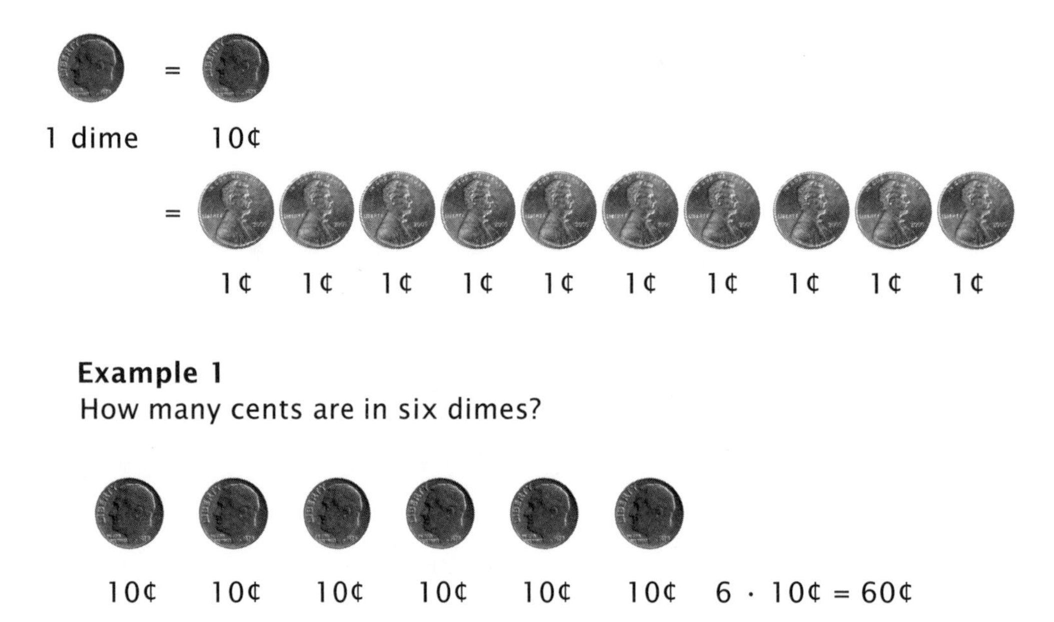

Example 1

How many cents are in six dimes?

10¢ 10¢ 10¢ 10¢ 10¢ 10¢ $6 \cdot 10¢ = 60¢$

We will be reviewing and using multiplication facts throughout the student workbook. Go to mathusee.com for more resources that may be used to review multiplication facts.

LESSON 6

Multiply by 5, 5¢ = 1 Nickel

After the two facts and ten facts have been mastered, we turn our attention to the five facts. This is a good time to teach odd and even numbers. ***Even numbers*** are multiples of two and end in 0, 2, 4, 6, or 8. ***Odd numbers*** end in 1, 3, 5, 7, or 9. An even number times five will end in zero. An odd number times five will end in five. Notice the pattern that emerges and reinforce this using the manipulatives.

Multiplying 5 by 1, 3, 5, or 7 yields products that end in 5 (5, 15, 25, and 35). Multiplying 5 by 2, 4, or 6 gives answers that end in 0 (10, 20, and 30).

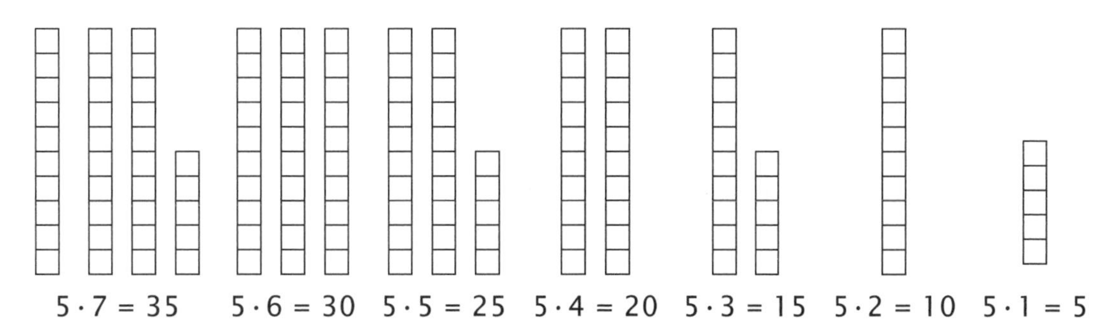

Another way to show this is on a number chart. Circling all of the five facts, or multiples of five, reveals an interesting pattern that corresponds to the block pattern shown above.

(0)	1	2	3	4	(5)	6	7	8	9
(10)	11	12	13	14	(15)	16	17	18	19
(20)	21	22	23	24	(25)	26	27	28	29
(30)	31	32	33	34	(35)	36	37	38	39
(40)	41	42	43	44	(45)	46	47	48	49

Since the student has learned the skip counting facts and is able to say them independently, he knows all of the multiples of five, or products of five. This is a good time to put the factors with the products. Say the factors slowly and then ask the student to say the products by skip counting. For example, you say "five counted one time" or "five times one," and the student says "five." You continue by saying "five times two," and the student says "ten." Proceed through all the facts sequentially. Here are the five facts with the corresponding products.

0	5	10	15	20	25	30	35	40	45	50
5×0	5×1	5×2	5×3	5×4	5×5	5×6	5×7	5×8	5×9	5×10
	↑					↑				↑

5 counted 1 time | 5 counted 6 times | 5 counted 10 times

Each five fact can be built in the shape of a rectangle. Whenever illustrating with the blocks, also write the problem and say it as you build. Notice that the drawing also illustrates the fact that 25 is five times greater than five.

5 counted five times is the same as 25;
5 times 5 equals 25; 5 over and 5 up is 25.

We can use the rectangles to emphasize the multiplication problems that correspond to the multiples of five. Here are a few examples.

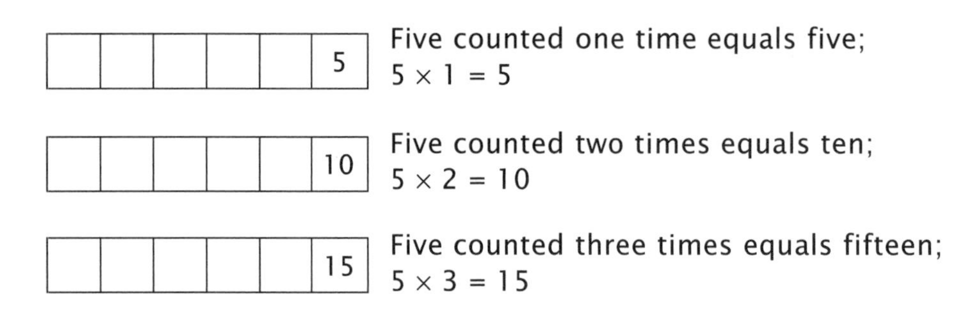

Another pattern I've observed that helps in learning the five facts builds on what we have already learned with the ten facts. There are two ways to use this pattern. In the first approach, recall that $10 \times 6 = 60$, and half of that is 30. Because 5 is half of 10, the answer to 5×6 is half of 60, or 30. We first multiplied by 10 (added a zero to the factor) and then found half of the product. You can build this to illustrate the concept. This method presumes that the student knows how to take a half of something. The picture below illustrates how this works.

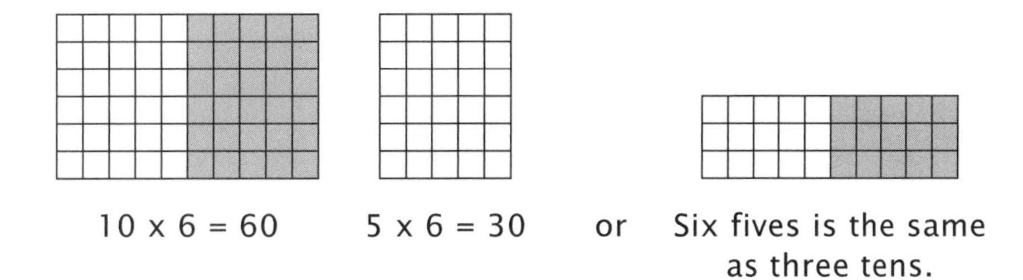

In the second approach to multiplying by five, we can divide the second factor in half and then multiply the result by 10. For 5 × 6, take half of 6, which is 3, and then multiply 3 by 10 (add a zero). This approach works for all even numbers.

Example 1

5 × 12 Take half of 12 and add a 0 to get 60.

or

5 × 12 Multiply 12 by 10 (add a 0) to get 120 and then divide the product in half to get 60.

0 × 0	0 × 1	0 × 2	0 × 3	0 × 4	0 × 5	0 × 6	0 × 7	0 × 8	0 × 9	0 × 10
1 × 0	1 × 1	1 × 2	1 × 3	1 × 4	1 × 5	1 × 6	1 × 7	1 × 8	1 × 9	1 × 10
2 × 0	2 × 1	2 × 2	2 × 3	2 × 4	2 × 5	2 × 6	2 × 7	2 × 8	2 × 9	2 × 10
3 × 0	3 × 1	3 × 2	3 × 3	3 × 4	3 × 5	3 × 6	3 × 7	3 × 8	3 × 9	3 × 10
4 × 0	4 × 1	4 × 2	4 × 3	4 × 4	4 × 5	4 × 6	4 × 7	4 × 8	4 × 9	4 × 10
5 × 0	5 × 1	5 × 2	5 × 3	5 × 4	5 × 5	5 × 6	5 × 7	5 × 8	5 × 9	5 × 10
6 × 0	6 × 1	6 × 2	6 × 3	6 × 4	6 × 5	6 × 6	6 × 7	6 × 8	6 × 9	6 × 10
7 × 0	7 × 1	7 × 2	7 × 3	7 × 4	7 × 5	7 × 6	7 × 7	7 × 8	7 × 9	7 × 10
8 × 0	8 × 1	8 × 2	8 × 3	8 × 4	8 × 5	8 × 6	8 × 7	8 × 8	8 × 9	8 × 10
9 × 0	9 × 1	9 × 2	9 × 3	9 × 4	9 × 5	9 × 6	9 × 7	9 × 8	9 × 9	9 × 10
10 × 0	10 × 1	10 × 2	10 × 3	10 × 4	10 × 5	10 × 6	10 × 7	10 × 8	10 × 9	10 × 10

5¢ = 1 Nickel

A good place to apply the five facts is with money. We know that 5¢ is the same as one nickel, so we can ask the student to find the number of cents in four nickels to apply 4×5. The answer is 20 cents, or 20¢.

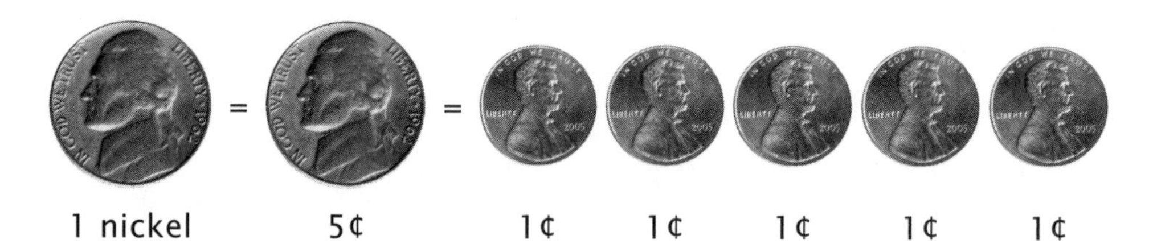

Example 2
How many cents are in four nickels?

$4 \cdot 5¢ = 20¢$

LESSON 7

Area of a Rectangle and a Square

The rectangle is the primary way we illustrate multiplication. The over dimension and the up dimension are the factors, and the area is the product. Multiplication of the factors gives the product. Up to now, we have always had squares inside the rectangles to show the area.

Now we will be applying our multiplication skills to find the area of rectangles when given the dimensions in units of measure such as inches (" or *in*) or feet (' or *ft*). Area is usually measured in square feet, square inches, etc. Use "square units" if a different unit of measure has not been specified. To help the students remember this, I use the word "squarea." This combines the two words "square" and "area."

Study the examples to make sure you understand how to find the area of a rectangle or a square. When in doubt, build the rectangle with the blocks to find the total number of square units.

Example 1
Find the area of the rectangle.

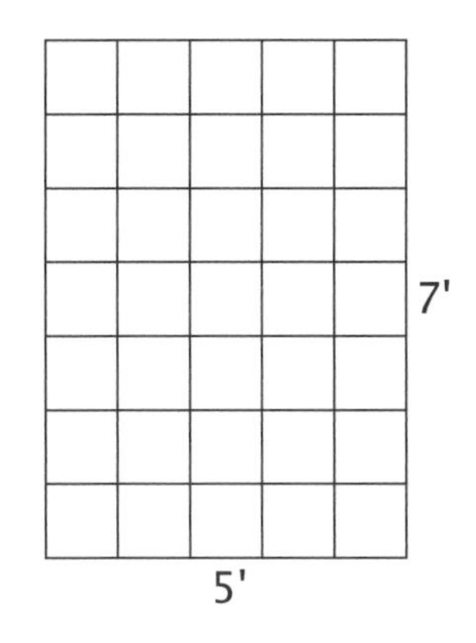

Area = (5 ft)(7 ft) = 35 square feet, or 35 sq ft
Check your answer by counting the squares.

Example 2
Find the area of the rectangle.

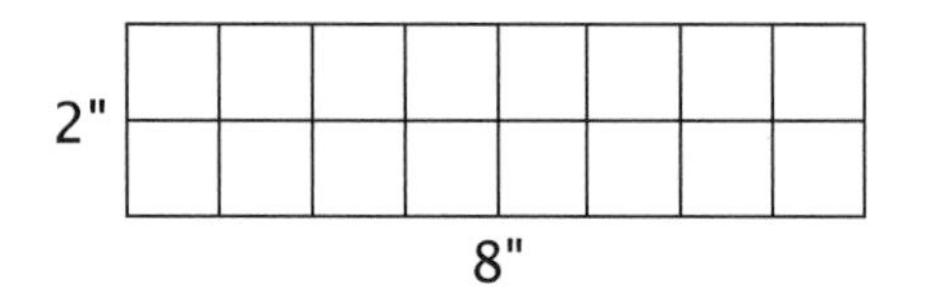

Area = (8 in)(2 in) = 16 square inches, or 16 sq in

Example 3
Find the area of the square. When a unit of measure is not specified, label your answer with the generic term *units*.

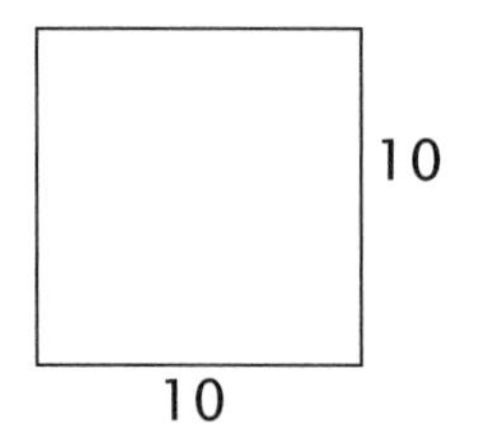

Area = (10 units)(10 units) = 100 square units, or 100 sq units

Example 4
Find the area of the square. Since the figure is a square, if we know the length of one side, we know the length of all the sides.

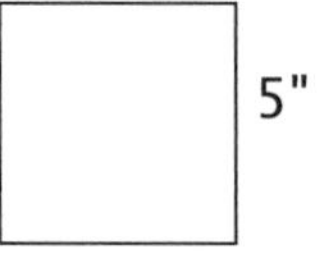

Area = (5 in)(5 in) = 25 square inches, or 25 sq in

Metric Measures and Meters

Most of the area problems in *Gamma* have the lengths of the sides given using feet, inches, or another U. S. customary unit. A few problems in this lesson are labeled with metric length units. Use a ruler to show students how centimeters compare to inches. Tell them that one hundred centimeters make one meter, which is a little longer than a yard. When you ask students to measure real objects, be sure to have them practice using centimeters as well as inches.

LESSON 8

Solve for an Unknown

While teaching the multiplication facts, we also teach solving for an unknown. There are three reasons why we introduce this topic now. Solving for an unknown reviews multiplication, introduces algebra concretely, and lays a solid foundation for division. This is not a light topic. Solving for an unknown does lay a foundation for the formal study of algebra, but that is still several years off. Most importantly, teaching this subject now will have an impact on learning single-digit division in *Delta*, the next level of Math-U-See. (See lesson 20 for more on solving for the unknown and division.) In multiplication you are given two factors, and you have to find the product. In division you are given the product and one factor, and you have to find the missing factor. This is exactly what we are doing in solving for an unknown factor.

How you say the equation is key to understanding what it means. 2G = 12 can be read as, "Two counted how many times is twelve?" or "What number counted two times is twelve?" This is because of the Commutative Property of Multiplication, which indicates that changing the order of the factors does not change the product. You might also say, "Two times what equals twelve?" Choose the way that is easiest for your student to understand. Example 1 illustrates this with the blocks. Study the examples until this important concept is understood.

Example 1
Solve for the unknown, or find the value of "G" in 2G = 12.

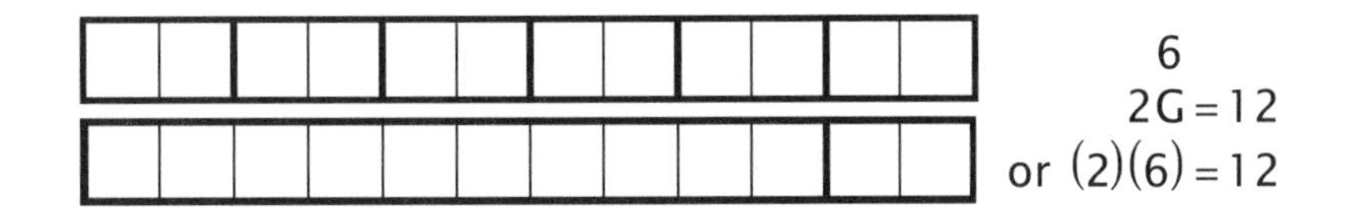

I solved $2G = 12$ as "How many twos can I count out of twelve?" or "Two times what equals twelve?" Using the blocks, you can see that there are six twos in twelve. Since $2 \times 6 = 12$, the missing factor must be 6. When you find the answer, simply write the 6 just above the G. Later on, in *Pre-Algebra*, we'll begin writing another line below the problem, with the solution written as $G = 6$.

Example 2

Solve for the unknown, or find the value of "X" in $5X = 15$.

$$\overset{3}{5X} = 15$$

$$\text{or } (5)(3) = 15$$

I solved it as "How many fives can I count out of fifteen?" or "Five times what equals fifteen?" Using the blocks, you can see that there are three fives in fifteen. Since $5 \times 3 = 15$, the missing factor must be 3. When you find the answer, simply write the 3 just above the X.

Classifying Quadrilaterals

A square is a rectangle whose sides all have the same length. Squares and rectangles are examples of ***quadrilaterals***. In fact, any closed figure with four straight sides is a quadrilateral. A ***rhombus*** is a quadrilateral whose sides all have the same length. It may or may not have square corners. A rhombus with square corners is commonly called a square. A rhombus without square corners is sometimes referred to as a diamond or kite shape.

LESSON 9

Skip Count by 9, Equivalent Fractions

In this lesson we are skip counting by nine. Use the techniques that you have used before to introduce and teach this important skill. For example, you could use skip counting by nine to find the number of starting players on several baseball teams.

Example 1

Skip count and write the numbers on the lines. Say each number out loud as you count and write. After filling in the blanks, write the numbers in the spaces provided beneath the figure. The solution is shown below.

								___
								___
								___

9 , ___ , ___

Solution

								9
								18
								27

9 , 18 , 27

Multiples in Equivalent Fractions

In lesson 3, one of the reasons cited for learning to skip count was the advantage it gives you when learning equivalent fractions. On the worksheets, we will give series of equivalent fractions and ask students to fill in the missing numerators and denominators by skip counting. Students do not need to know what a fraction is to find the missing numbers, but, if you wish, you can refer to lesson 13, where the concept of equivalent fractions is explained.

In example 2, we see the multiples of 2/5.

Example 2

Find the missing multiples of 2 and 5 in the equivalent fractions.

$$\frac{2}{5} = \frac{4}{\quad} = \frac{6}{15} = \frac{\quad}{\quad} = \frac{\quad}{25} = \frac{12}{\quad} = \frac{\quad}{35} = \frac{\quad}{\quad} = \frac{\quad}{\quad} = \frac{20}{50}$$

Solution

$$\frac{2}{5} = \frac{4}{10} = \frac{6}{15} = \frac{8}{20} = \frac{10}{25} = \frac{12}{30} = \frac{14}{35} = \frac{16}{40} = \frac{18}{45} = \frac{20}{50}$$

LESSON 10

Multiply by 9

Multiplication by nine has a unique pattern. Begin with nine green units to represent 1×9 or 9. To show 2×9, you need to add nine more, but the units place is full. Instead, add a ten block, which is one too many, and take one block away from the units place. By adding one ten to the tens place and subtracting one unit from the units place, you have added nine. The answer for 2×9 is 18.

Adding another ten to the tens place and taking another unit from the units place gives me three nines, or 3×9, which is 27. Continuing on by adding ten and subtracting one each time, I proceed through all of the multiples of nine: 9–18–27–36–45–54–63–72–81–90. Because I took away a unit whenever I added a ten, I always added nine blocks. This reveals an interesting pattern. Add the individual digits of each answer. For 18, $1 + 8 = 9$, for 27, $2 + 7 = 9$, and for 36, $3 + 6 = 9$. We can tell whether a number is a multiple of nine by adding the digits and seeing if they will add up to nine or a multiple of nine. For example, take 144 and add the digits: $1 + 4 + 4 = 9$. This means that 144 should be a multiple of nine, and, sure enough, $9 \times 16 = 144$.

We can relate this to the 10 facts. We know 10×6 is 60, so 9×6 will be a little less than 60. It is 50-something, or 5___. What plus 5 makes 9? The answer is 4. Therefore, $9 \times 6 = 54$. Do several of these to discover a pattern. We took one less than six for the number in the tens place and then found the number that could be added to five to make nine. The final answer is 54.

9×8 The tens place is one less than 8, which is 7, and $7 + 2 = 9$, so the answer is 72.

9×4 The tens place is one less than 4, which is 3, and $3 + 6 = 9$, so the answer is 36.

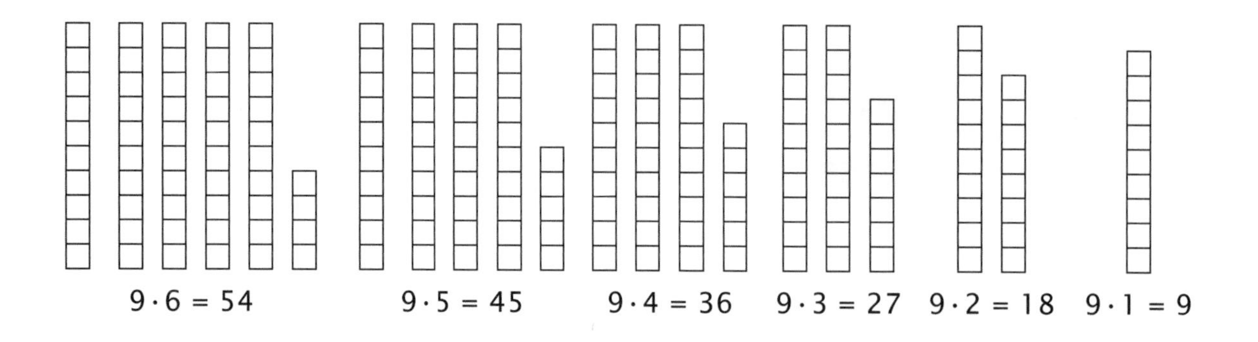

Another way to show this is on a number chart. Circling all of the nine facts, or multiples of nine, reveals the pattern that corresponds to the blocks above.

0	1	2	3	4	5	6	7	8	(9)
10	11	12	13	14	15	16	17	(18)	19
20	21	22	23	24	25	26	(27)	28	29
30	31	32	33	34	35	(36)	37	38	39
40	41	42	43	44	(45)	46	47	48	49
50	51	52	53	(54)	55	56	57	58	59
60	61	62	(63)	64	65	66	67	68	69
70	71	(72)	73	74	75	76	77	78	79
80	(81)	82	83	84	85	86	87	88	89
(90)									

Each nine fact can be built in the shape of a rectangle. Whenever illustrating with the blocks, also write the problem and say it as you build. Notice that the drawing also illustrates the fact that 54 is nine times greater than six.

Nine counted six times is the same as 54;
9 times 6 equals 54; 9 over and 6 up is 54.

We can use skip counting to emphasize the multiplication problems that correspond to the multiples of nine. Here are a few examples.

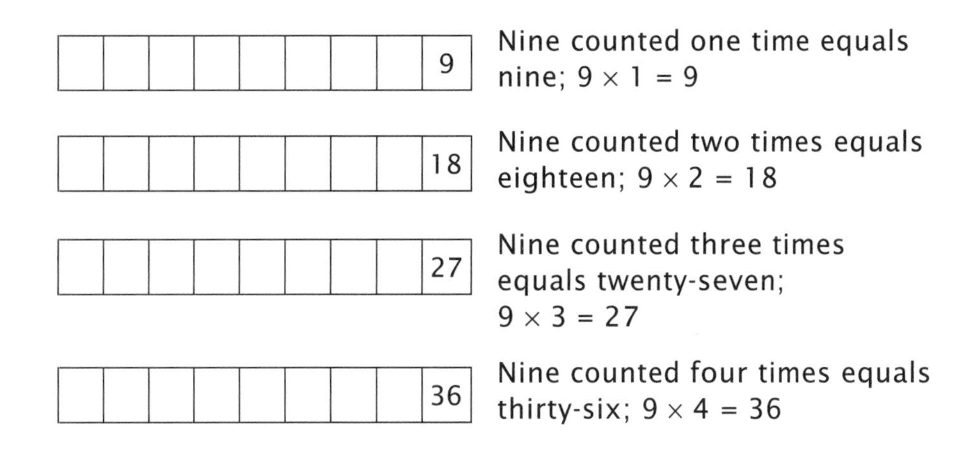

Skip counting by nine is the first step. After this is accomplished, say the factors slowly and ask the student to say the products. For example, you say "nine counted one time" or "nine times one," and the student says "nine." Continue by saying "nine times two" and have the student say "18." Proceed through all the facts sequentially.

Here are the nine facts with the corresponding products.

$$\frac{0}{(9)(0)} \quad \frac{9}{(9)(1)} \quad \frac{18}{(9)(2)} \quad \frac{27}{(9)(3)} \quad \frac{36}{(9)(4)} \quad \frac{45}{(9)(5)} \quad \frac{54}{(9)(6)} \quad \frac{63}{(9)(7)} \quad \frac{72}{(9)(8)} \quad \frac{81}{(9)(9)} \quad \frac{90}{(9)(10)}$$

↑ 9 counted 0 time ↑ 9 counted 5 times ↑ 9 counted 8 times

0 × 0	0 × 1	0 × 2	0 × 3	0 × 4	0 × 5	0 × 6	0 × 7	0 × 8	0 × 9	0 × 10
1 × 0	1 × 1	1 × 2	1 × 3	1 × 4	1 × 5	1 × 6	1 × 7	1 × 8	1 × 9	1 × 10
2 × 0	2 × 1	2 × 2	2 × 3	2 × 4	2 × 5	2 × 6	2 × 7	2 × 8	2 × 9	2 × 10
3 × 0	3 × 1	3 × 2	3 × 3	3 × 4	3 × 5	3 × 6	3 × 7	3 × 8	3 × 9	3 × 10
4 × 0	4 × 1	4 × 2	4 × 3	4 × 4	4 × 5	4 × 6	4 × 7	4 × 8	4 × 9	4 × 10
5 × 0	5 × 1	5 × 2	5 × 3	5 × 4	5 × 5	5 × 6	5 × 7	5 × 8	5 × 9	5 × 10
6 × 0	6 × 1	6 × 2	6 × 3	6 × 4	6 × 5	6 × 6	6 × 7	6 × 8	6 × 9	6 × 10
7 × 0	7 × 1	7 × 2	7 × 3	7 × 4	7 × 5	7 × 6	7 × 7	7 × 8	7 × 9	7 × 10
8 × 0	8 × 1	8 × 2	8 × 3	8 × 4	8 × 5	8 × 6	8 × 7	8 × 8	8 × 9	8 × 10
9 × 0	9 × 1	9 × 2	9 × 3	9 × 4	9 × 5	9 × 6	9 × 7	9 × 8	9 × 9	9 × 10 ←
10 × 0	10 × 1	10 × 2	10 × 3	10 × 4	10 × 5	10 × 6	10 × 7	10 × 8	10 × 9	10 × 10
									↑	

Adding and Subtracting Time

Telling time was taught in *Primer*, *Alpha*, and *Beta*. When you are ready to teach addition and subtraction of time, begin by having the student add increments of five minutes, skip counting by fives around the face of the clock. In effect, the student is using the clock face as a circular number line.

When this is understood, move on to using regular addition to add minutes that are not multiplies of five and then teach subtraction in the same way. For now, avoid problems that require regrouping, as this requires regrouping 60, rather than 10 or 100.

LESSON 11

Skip Count by 3

Before we learn how to multiply by three, we must master skip counting by three. If the student learns skip counting by singing songs, make sure he can also say the numbers without the aid of music. Some objects you can skip count by three are the sides of a triangle or the wheels on a tricycle.

Example 1

Skip count and write the numbers on the lines. Say each number out loud as you count and write. After filling in the rectangle, write the numbers in the spaces provided beside the figure.

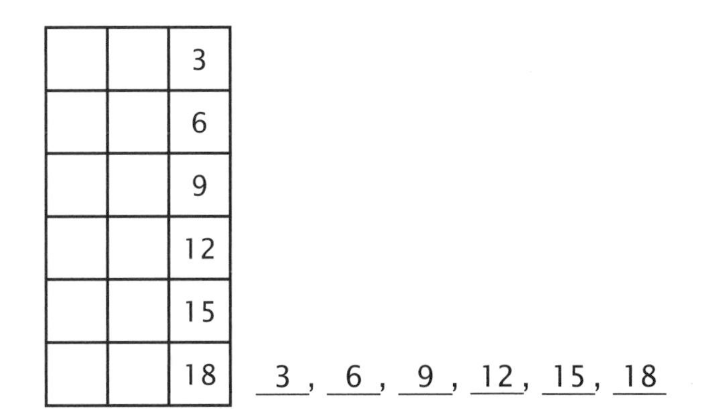

Notice that this rectangle illustrates a specific multiplication problem: $3 \times 6 = 18$

On the next page are examples of how to practice this new skill by counting the sides of a triangle or the dots on one side of a domino.

Example 2
How many sides are on four triangles?

"3-6-9-12" There are 12 sides on four triangles.

Example 3
How many dots are on six sides?

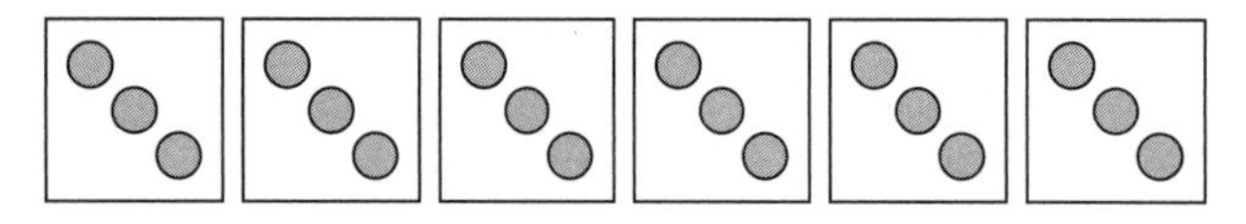

"3-6-9-12-15-18" There are 18 dots on six sides.

LESSON 12

Multiply by 3, 3 Feet = 1 Yard

1 Tablespoon = 3 Teaspoons

So far we have learned the 0, 1, 2, 5, 9, and 10 facts. The chart at the end of the lesson shows that we have already learned some of the 3 facts. These facts are left: 3×3, 3×4, 3×6, 3×7, and 3×8. Be encouraged; we are over halfway finished.

Use all of the techniques you have previously used for learning the facts. You can also simply memorize them. It is very important that the students have their facts mastered before moving any further. All of the strategies and patterns are meant to be aids to the learning process; they are not ends in themselves.

Notice the products of the three facts. The digits in the products add up to three or a multiple of three. For example, for 21, $2 + 1 = 3$; for 24, $2 + 4 = 6$; and for 27, $2 + 7 = 9$. This is similar to the pattern for the nines.

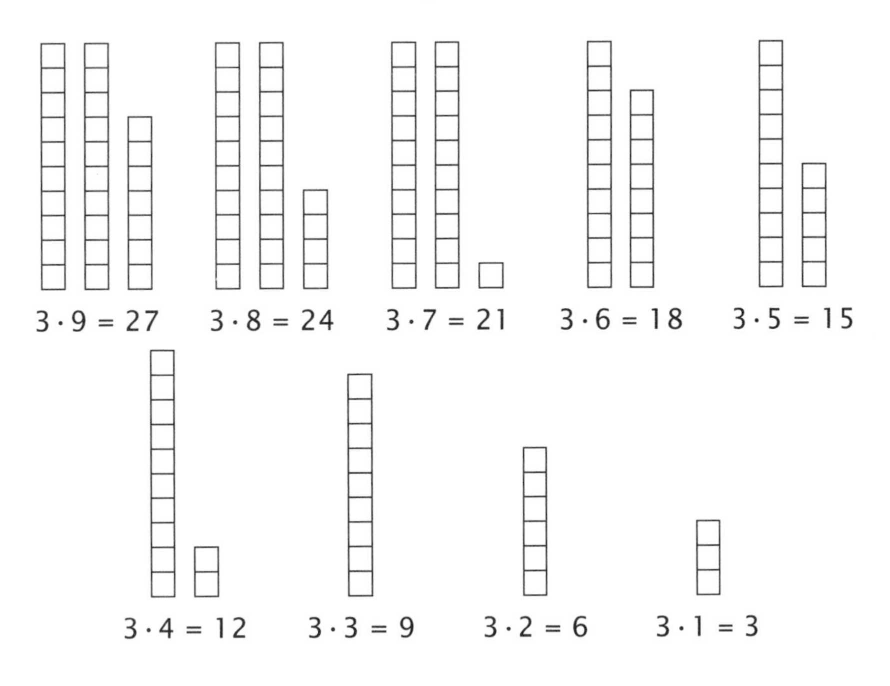

Another way to show this is on a number chart. Circling all of the three facts, or multiples of three, reveals an interesting pattern that corresponds to the blocks on the previous page.

0	1	2	(3)	4	5	(6)	7	8	(9)
10	11	(12)	13	14	(15)	16	17	(18)	19
20	(21)	22	23	(24)	25	26	(27)	28	29

Since the student has learned the facts by skip counting, he or she knows all of the multiples of three, or products of three. Say the factors slowly and ask the student to say the products by skip counting. You say, "three counted one time" or "three times one," and the student says "three." You continue by saying, "three times two," and the student says "six." Proceed through all the facts sequentially.

Here are the 3 facts with the corresponding products.

0	3	6	9	12	15	18	21	24	27	30
3×0	3×1	3×2	3×3	3×4	3×5	3×6	3×7	3×8	3×9	3×10
		↑				↑				↑
		3 counted 2 times				3 counted 6 times				3 counted 10 times

Each three fact can be built in the shape of a rectangle. Whenever illustrating with the blocks, also write the problem and say it as you build. The drawing also illustrates the fact that 12 is three times greater than four.

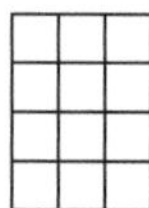

3 counted four times is the same as 12;
3 times 4 equals 12; 3 over and 4 up is 12.

We can use the rectangles on the worksheets and skip counting to emphasize the multiplication problems that correspond to the multiples of three. Some examples are shown below.

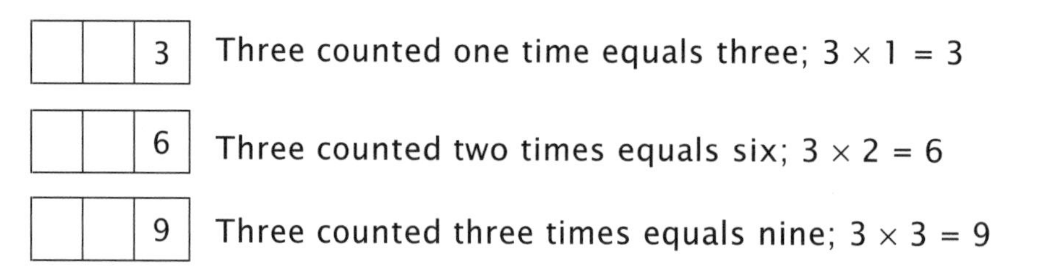

3 Feet = 1 Yard

Show the students a one-foot ruler and a three-foot yardstick. Hold up the ruler and explain that this is one foot long. A foot measures length. Measure the height of some people and see how tall they are. Measure the distance to the ceiling or the width of a desk or table. The symbol for feet is '. Four feet can be written as as 4', and six feet can be written as 6'.

Introduce the yardstick and, using the ruler, show that it is the same length as three rulers; therefore, it is three feet long. Measure objects with the yardstick, such as the length of the room. If the students are interested in football, show them five yards (the distance between lines on a football field) and then show 10 yards (what is needed for a first down). If they like track, mention the 100-yard dash and the 50-yard dash.

After they are familiar with the yardstick, tell the students that the man who wrote this book is around two yards tall and ask how many feet that represents. If there are three feet in one yard, there are three plus three, or six feet, in two yards. Changing yards to feet is an appropriate place to apply multiplying by threes.

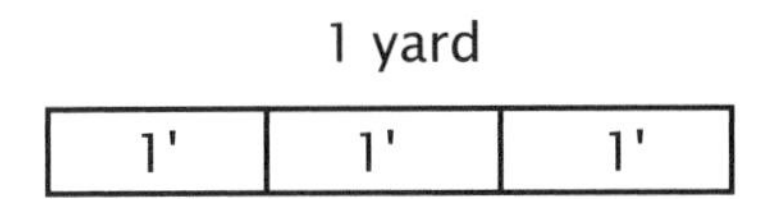

Example 1

How many feet are in two yards?

2 yards = 2 × 3 feet (or 1 yard) = 6 feet

Liquid Measure: Teaspoons and Tablespoons

Begin with a measuring teaspoon and a measuring tablespoon. Fill up the teaspoon and empty it into the tablespoon. Do this three times. The student sees that one tablespoon equals three teaspoons.

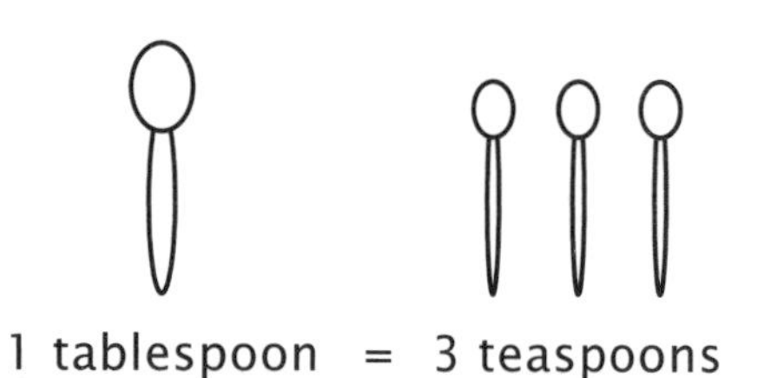

Abbreviations:
Tablespoon = Tbsp
teaspoon = tsp

Example 2

How many teaspoons are in four tablespoons?

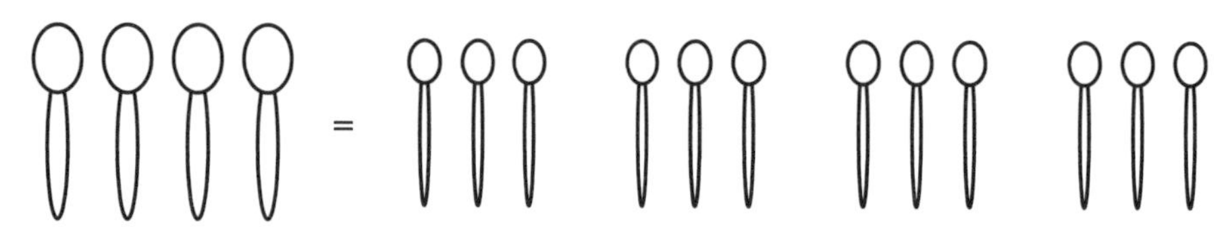

4 Tbsp = 4 × 3 (tsp in a Tbsp) = 12 tsp

0 × 0	0 × 1	0 × 2	0 × 3	0 × 4	0 × 5	0 × 6	0 × 7	0 × 8	0 × 9	0 × 10
1 × 0	1 × 1	1 × 2	1 × 3	1 × 4	1 × 5	1 × 6	1 × 7	1 × 8	1 × 9	1 × 10
2 × 0	2 × 1	2 × 2	2 × 3	2 × 4	2 × 5	2 × 6	2 × 7	2 × 8	2 × 9	2 × 10
3 × 0	3 × 1	3 × 2	3 × 3	3 × 4	3 × 5	3 × 6	3 × 7	3 × 8	3 × 9	3 × 10
4 × 0	4 × 1	4 × 2	4 × 3	4 × 4	4 × 5	4 × 6	4 × 7	4 × 8	4 × 9	4 × 10
5 × 0	5 × 1	5 × 2	5 × 3	5 × 4	5 × 5	5 × 6	5 × 7	5 × 8	5 × 9	5 × 10
6 × 0	6 × 1	6 × 2	6 × 3	6 × 4	6 × 5	6 × 6	6 × 7	6 × 8	6 × 9	6 × 10
7 × 0	7 × 1	7 × 2	7 × 3	7 × 4	7 × 5	7 × 6	7 × 7	7 × 8	7 × 9	7 × 10
8 × 0	8 × 1	8 × 2	8 × 3	8 × 4	8 × 5	8 × 6	8 × 7	8 × 8	8 × 9	8 × 10
9 × 0	9 × 1	9 × 2	9 × 3	9 × 4	9 × 5	9 × 6	9 × 7	9 × 8	9 × 9	9 × 10
10 × 0	10 × 1	10 × 2	10 × 3	10 × 4	10 × 5	10 × 6	10 × 7	10 × 8	10 × 9	10 × 10

LESSON 13

Skip Count by 6, Equivalent Fractions

Skip counting by six will help us learn to multiply by six. Even after we learn our facts, skip counting can still aid us. There are times when I have seen a student forget that $6 \times 6 = 36$. Instead of panicking, a student who knows how to skip count and understands that skip counting is fast adding of the same number can figure it out. He or she might say, "I know that $5 \times 6 = 30$, so 6×6 must be 6 more than that, or 36." Reinforce this as you learn these facts in order. If the student learns them by singing songs, make sure he can also skip count without the aid of the music.

Ideas for real-life application are the sides of a hexagon, the legs of an insect, a half dozen, the walls in a honeycomb, and the dots on a die (singular for dice).

Example 1

Skip count and write the numbers on the lines. Say each number out loud as you count and write. After filling in the rectangle, write the numbers in the spaces provided beneath the figure.

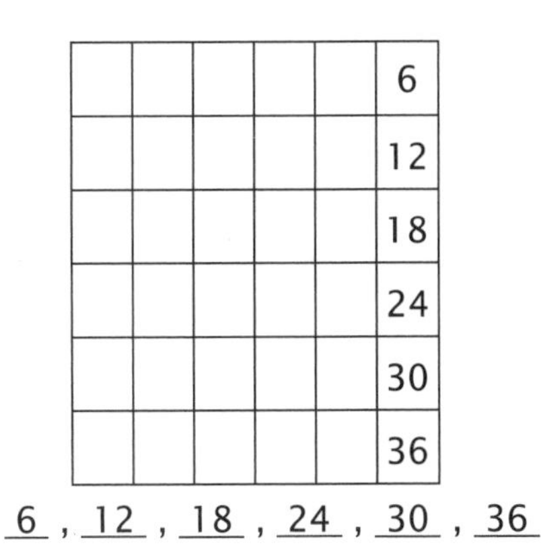

Notice that this rectangle illustrates a specific problem: $6 \times 6 = 36$

Fractions

For the last few lessons, we have been using multiples or skip counting to make equivalent fractions. In this lesson, we will show what a fraction and an equivalent fraction are using rectangles. A ***fraction*** has a numerator and a denominator. The ***denominator*** is the number that represents how many total parts there are. It is written on the bottom of the fraction. The ***numerator,*** or "numberator," is the number that represents how many of the total parts are shaded. It is written on the top of the fraction.

In the next few examples, first count the total number of parts and write that number on the bottom. Then count how many are shaded and write that number on the top.

Example 2
Find the denominator and numerator shown by the rectangle.

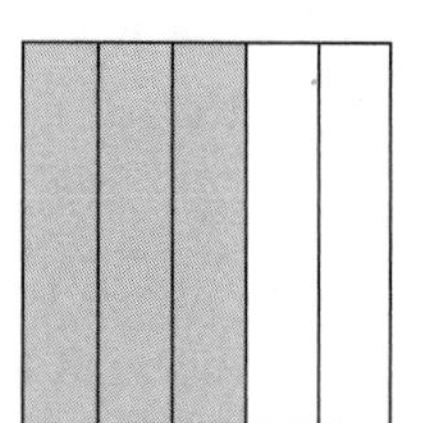

$$\frac{\text{numerator}}{\text{denominator}} = \frac{3}{5} = \frac{\text{how many}}{\text{total parts}}$$

Example 3
Find the denominator and numerator shown by the rectangle.

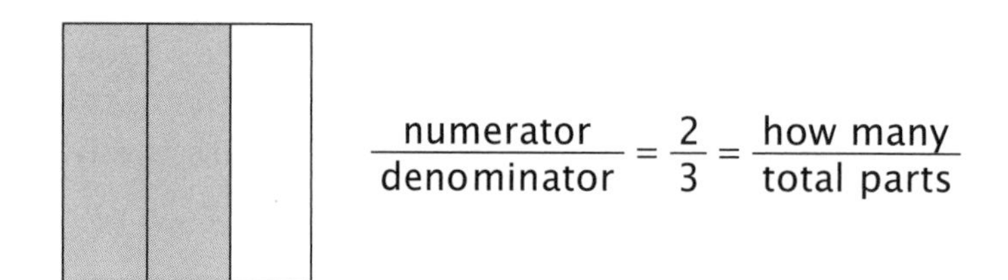

NOTE: The teaching on the DVD uses the overlays, but the student shouldn't need to have his own set of overlays to solve the problems. You can wait until the book on fractions before purchasing them. At this level, cooking and other real-life applications of fractions are recommended as illustrations.

Equivalent Fractions

An ***equivalent fraction*** is a fraction with the same amount but more pieces. Think of Example 3 as being a cake that was cut into three pieces. One of the pieces was eaten. There are two pieces left, but several people come to eat the cake, so they have to divide it into more pieces. It is still the same amount of cake (2/3), but there are more pieces, as shown in Example 4. Notice that the "numberator" and denominator of equivalent fractions are the multiples that we find as a result of skip counting.

Example 4

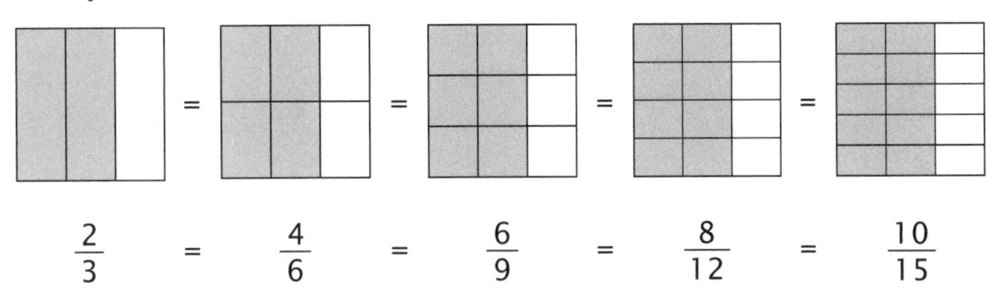

$$\frac{2}{3} = \frac{4}{6} = \frac{6}{9} = \frac{8}{12} = \frac{10}{15}$$

Example 5

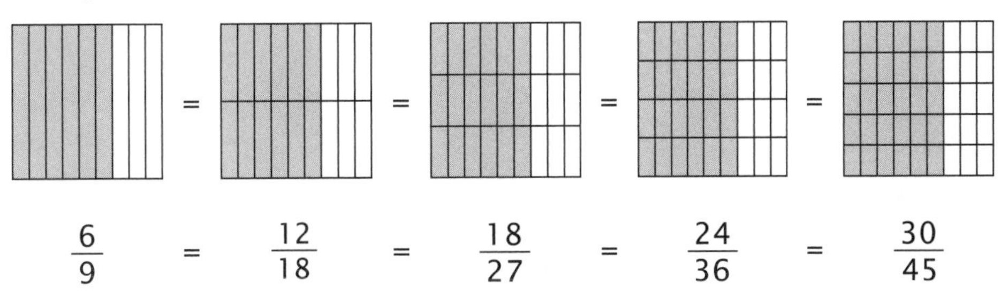

$$\frac{6}{9} = \frac{12}{18} = \frac{18}{27} = \frac{24}{36} = \frac{30}{45}$$

Example 6

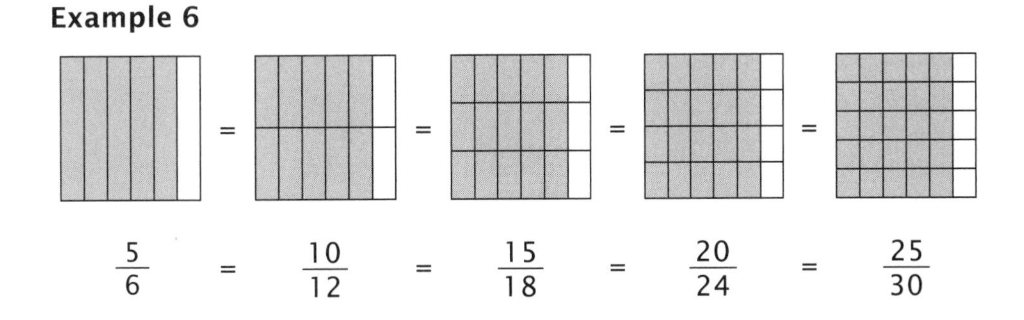

$$\frac{5}{6} = \frac{10}{12} = \frac{15}{18} = \frac{20}{24} = \frac{25}{30}$$

LESSON 14

Multiply by 6

We have only four of the six facts left to learn: 6×4, 6×6, 6×7, and 6×8. There are a few patterns that emerge as we study the six facts. Since six is a multiple of three, the digits in the answers will add up to a multiple of three. Since six is a multiple of two, all the answers will be even. Recently, someone noticed that when multiplying six by an even number, the answer always ends with that same number.

Using this pattern, take half of the multiplier for the tens place and then add the multiplier. For 6×4, take half of the number four, which is two, for the tens place, and then add the number four. The answer is 24. The pattern starts with knowing the fives. When we multiplied by five, we took half of the multiplier and then added a zero. Multiplying 5×4 gives us 20. Since six is one more than five, we must also take the multiplier times 1 to find the units value. To show this, consider that 6 is 5 + 1 and look at the picture below.

Try this with the other even multipliers. If this is confusing, don't worry about it and simply employ whatever strategy you used previously to learn the six facts.

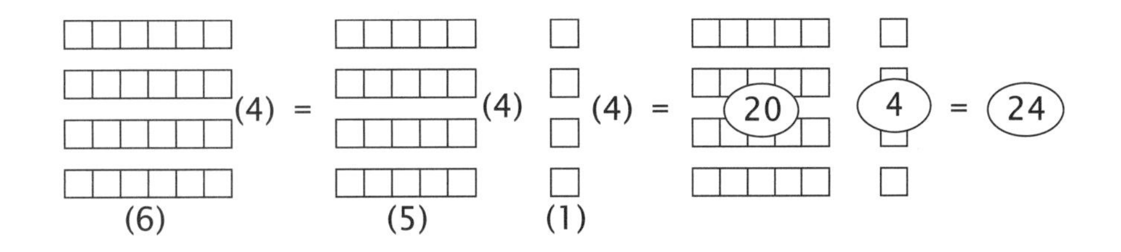

We can use the rectangles on the worksheets to emphasize the multiplication problems that correspond to the multiples of six. There is an example of this on the next page.

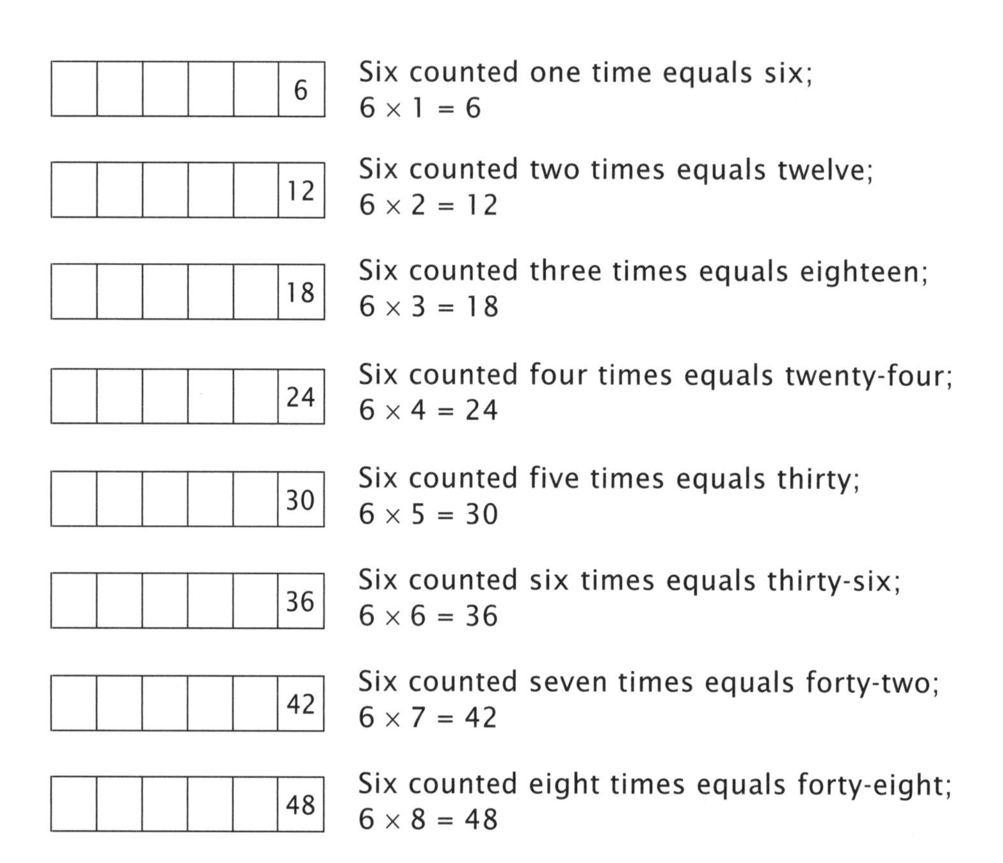

Six counted one time equals six;
$6 \times 1 = 6$

Six counted two times equals twelve;
$6 \times 2 = 12$

Six counted three times equals eighteen;
$6 \times 3 = 18$

Six counted four times equals twenty-four;
$6 \times 4 = 24$

Six counted five times equals thirty;
$6 \times 5 = 30$

Six counted six times equals thirty-six;
$6 \times 6 = 36$

Six counted seven times equals forty-two;
$6 \times 7 = 42$

Six counted eight times equals forty-eight;
$6 \times 8 = 48$

0 × 0	0 × 1	0 × 2	0 × 3	0 × 4	0 × 5	0 × 6	0 × 7	0 × 8	0 × 9	0 × 10
1 × 0	1 × 1	1 × 2	1 × 3	1 × 4	1 × 5	1 × 6	1 × 7	1 × 8	1 × 9	1 × 10
2 × 0	2 × 1	2 × 2	2 × 3	2 × 4	2 × 5	2 × 6	2 × 7	2 × 8	2 × 9	2 × 10
3 × 0	3 × 1	3 × 2	3 × 3	3 × 4	3 × 5	3 × 6	3 × 7	3 × 8	3 × 9	3 × 10
4 × 0	4 × 1	4 × 2	4 × 3	4 × 4	4 × 5	4 × 6	4 × 7	4 × 8	4 × 9	4 × 10
5 × 0	5 × 1	5 × 2	5 × 3	5 × 4	5 × 5	5 × 6	5 × 7	5 × 8	5 × 9	5 × 10
6 × 0	6 × 1	6 × 2	6 × 3	6 × 4	6 × 5	6 × 6	6 × 7	6 × 8	6 × 9	6 × 10 ←
7 × 0	7 × 1	7 × 2	7 × 3	7 × 4	7 × 5	7 × 6	7 × 7	7 × 8	7 × 9	7 × 10
8 × 0	8 × 1	8 × 2	8 × 3	8 × 4	8 × 5	8 × 6	8 × 7	8 × 8	8 × 9	8 × 10
9 × 0	9 × 1	9 × 2	9 × 3	9 × 4	9 × 5	9 × 6	9 × 7	9 × 8	9 × 9	9 × 10
10 × 0	10 × 1	10 × 2	10 × 3	10 × 4	10 × 5	10 × 6 ↑	10 × 7	10 × 8	10 × 9	10 × 10

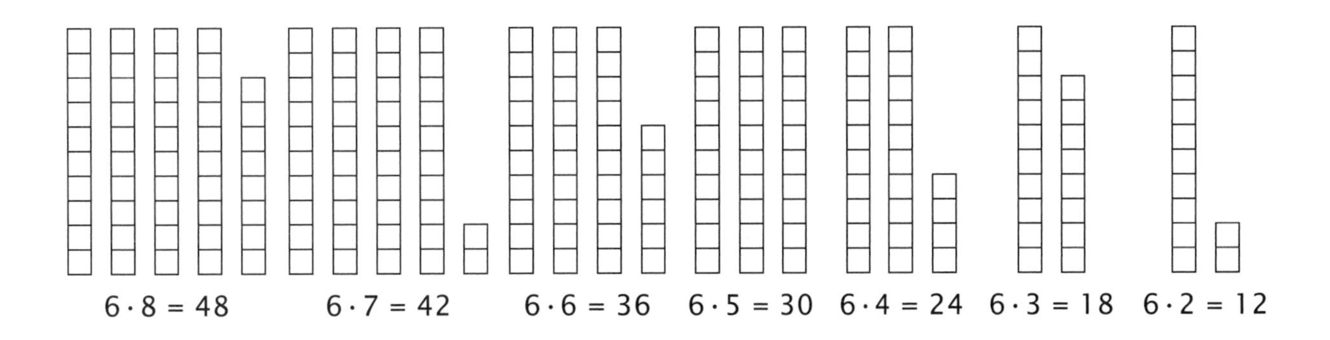

Another way to show this is on a number chart. Circling all of the six facts, or multiples of six, reveals an interesting pattern that corresponds to the blocks on the previous page.

0	1	2	3	4	5	6	7	8	9
10	11	12	13	14	15	16	17	18	19
20	21	22	23	24	25	26	27	28	29
30	31	32	33	34	35	36	37	38	39
40	41	42	43	44	45	46	47	48	49
50	51	52	53	54	55	56	57	58	59
60	61	62	63	64	65	66	67	68	69

Once the student has learned the skip counting facts and is able to say them without music, he or she knows all of the multiples of six or products of six. This is a good time to put the factors with the product. Say the factors slowly and ask the student to say the products by skip counting. For example, you say, "six counted one time" or "six times one," and the student says "six." Continue by saying "six times two" and having the student say "twelve." Proceed through all the six facts sequentially.

Here are the six facts with the corresponding products.

0	6	12	18	24	30	36	42	48	54	60
6×0	6×1	6×2	6×3	6×4	6×5	6×6	6×7	6×8	6×9	6×10
			↑			↑			↑	

6 counted 3 times   6 counted 6 times   6 counted 9 times

Each six fact can be built in the shape of a rectangle. Whenever illustrating with the blocks, also write the problem and say it as you build. The drawing also illustrates the fact that 36 is six times greater than six.

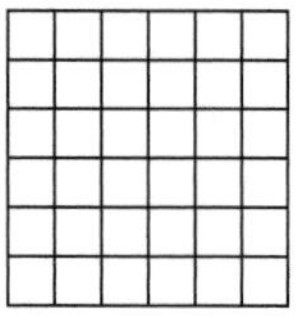

6 counted six times is the same as 36; 6 times 6 equals 36; 6 over and 6 up is 36.

LESSON 15

Skip Count by 4, 4 Quarts = 1 Gallon

Quarts and gallons are United States customary units for measuring the volume of liquids. There are four quarts in one gallon. After we know how to skip count or fast add by four, we can apply this to counting the number of quarts in several gallons. If you have three gallons, you have 4–8–12 quarts. There are illustrations of this on the next page.

Example 1

Skip count and write the numbers on the lines. Say each number out loud as you count and write. After filling in the rectangle, write the numbers in the spaces provided beneath the figure.

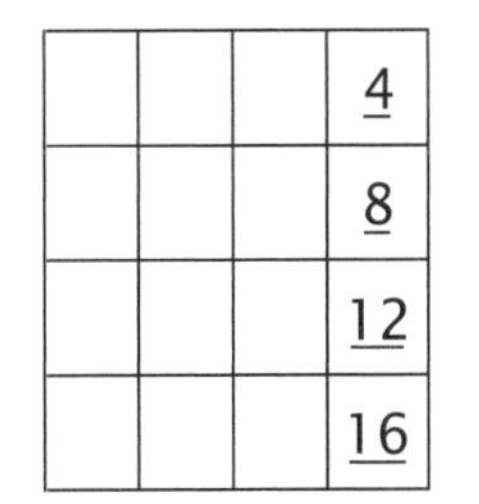

Notice that this rectangle illustrates a specific multiplication problem: $4 \times 4 = 16$

4 , 8 , 12 , 16

4 Quarts = 1 Gallon

If you can illustrate this with liquid, begin with four one-quart containers and one one-gallon container. Fill up the one-gallon container and empty it into the four one-quart containers. The student can see that one gallon equals four quarts. You can also show the converse by emptying the four quarts into the one gallon.

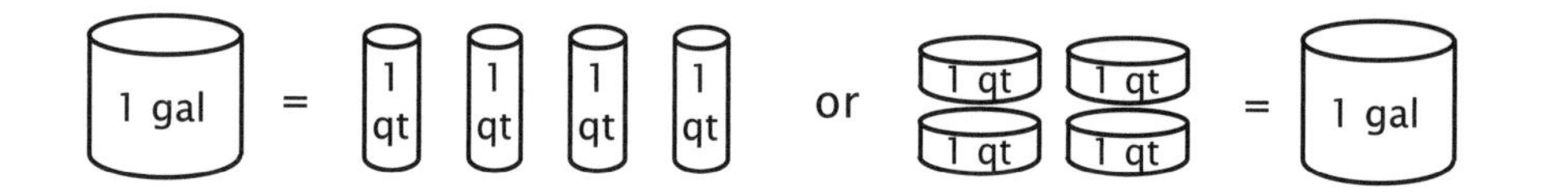

Example 2

How many quarts are in two gallons?

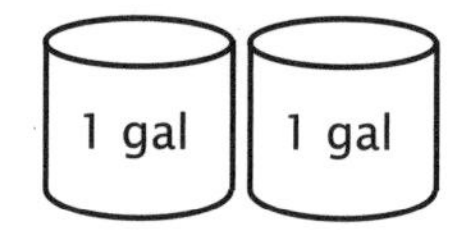

"4-8."

There are eight quarts in two gallons.

Example 3

How many quarts are in four gallons?

"4-8-12-16."

There are 16 quarts in four gallons.

Metric Measures and Liters

Most packages in the United States are labeled with both U. S. customary units (such as quarts and gallons) and metric units. A common metric unit for measuring volume is a ***liter***, which is slightly more than a U. S. quart. A liter is made up of 1000 ***milliliters***. One of the green Math-U-See unit blocks would hold about a milliliter of water.

At this level, we are not focusing on converting from one metric unit to another or from metric to customary units. However, students should be familiar with the words liter and milliliter and be able to estimate the volume of objects. Continue to use real-life examples for teaching all kinds of measures.

LESSON 16

Multiply by 4, 4 Quarters = 1 Dollar

The four facts we haven't studied yet are 4 × 4, 4 × 7, and 4 × 8. Learn these and review the ones you already know. Notice that the multiples of four are all even numbers and that they can all be broken down into two equal addends.

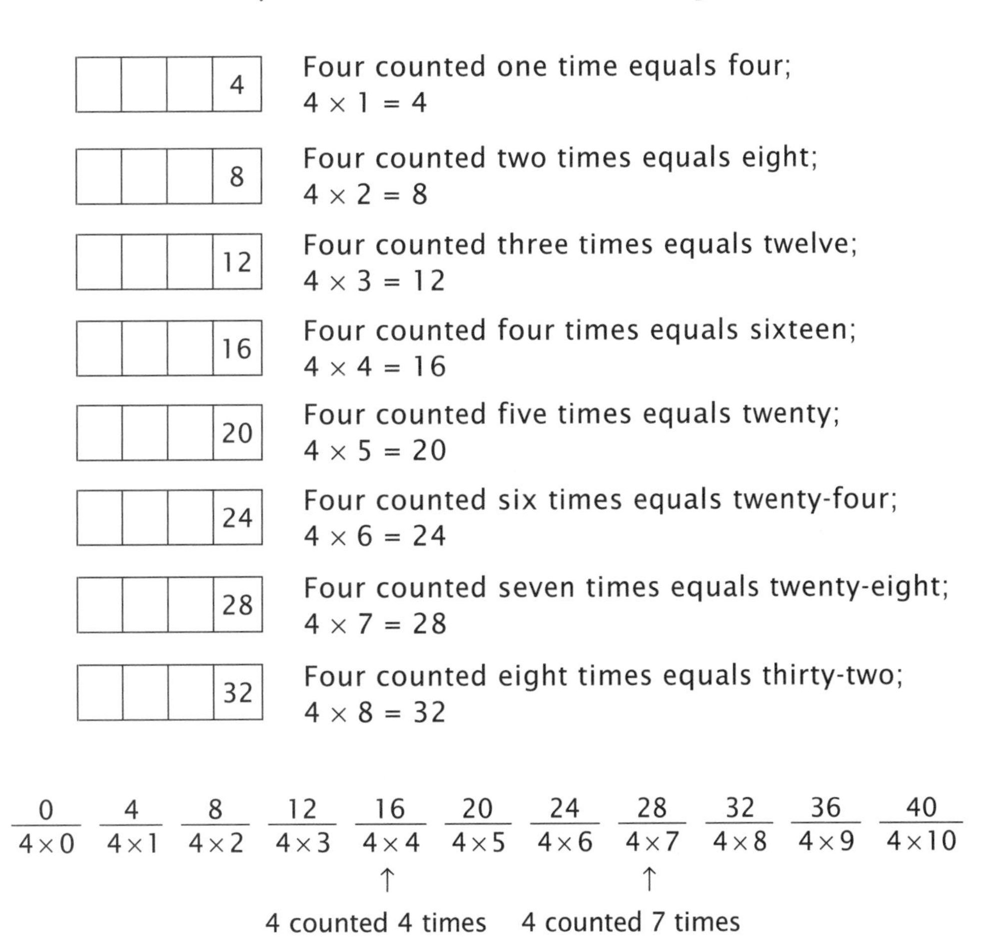

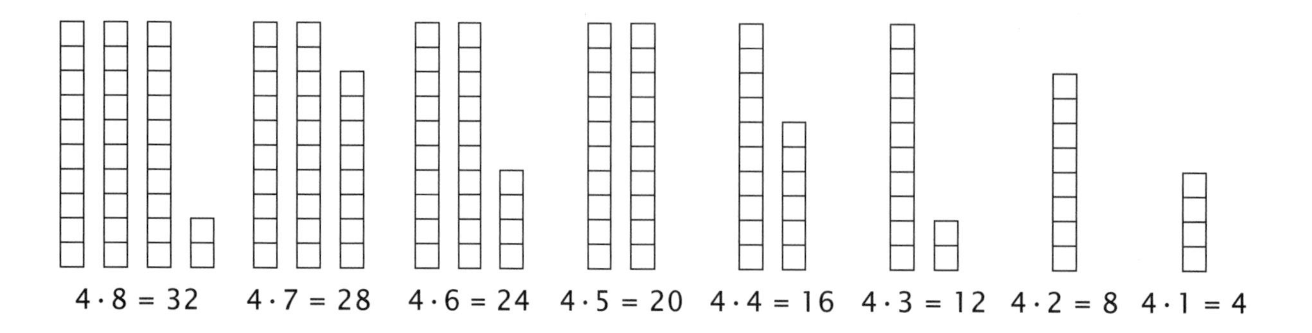

Another way to show this is on a number chart. Circling all of the four facts, or multiples of four, reveals an interesting pattern that corresponds to the blocks shown above.

0	1	2	3	(4)	5	6	7	(8)	9
10	11	(12)	13	14	15	(16)	17	18	19
(20)	21	22	23	(24)	25	26	27	(28)	29
30	31	(32)	33	34	35	(36)	37	38	39
(40)									

1 Quarter = 25¢ and 4 Quarters = 1 Dollar

One quarter has the same value as 25 cents. We can write it as 25¢. With the blocks, show one quarter by holding up two ten bars and one five bar.

Using the blocks, discover how many quarters are in one dollar (100 block). Then hold up a dollar bill in one hand and four quarters in the other to show that these are the same amount.

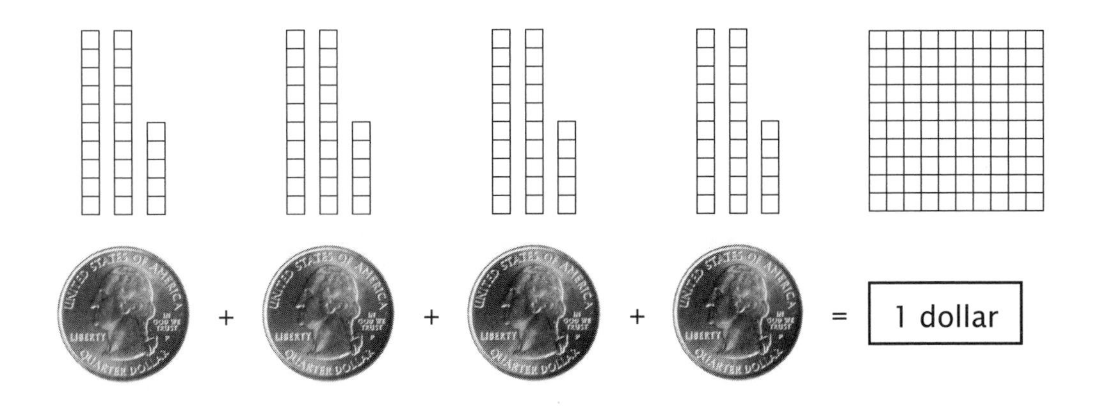

four quarters = one dollar

Example 1
How many quarters are in three dollars?

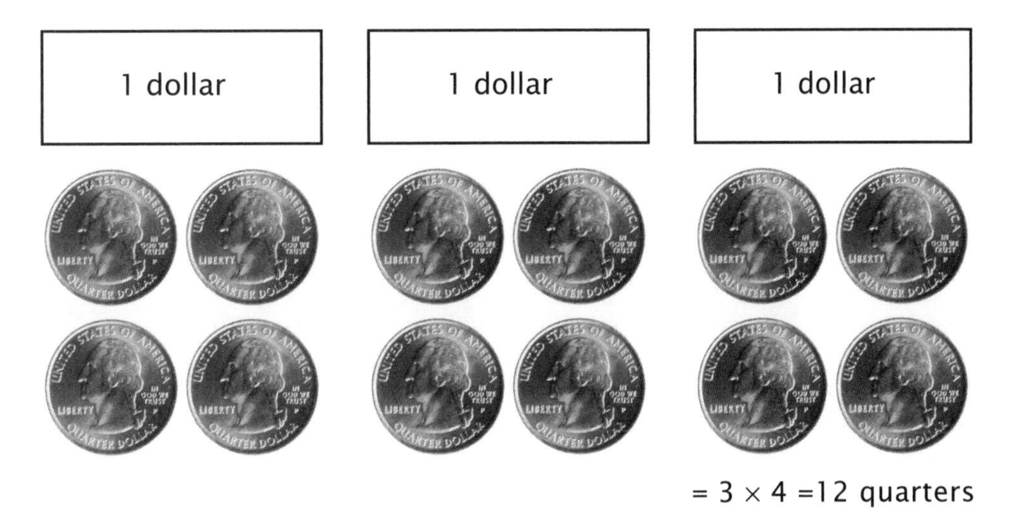

$= 3 \times 4 = 12$ quarters

Each four fact can be built in the shape of a rectangle. Whenever illustrating with the blocks, also write the problem and say it as you build. The drawing below illustrates the fact that 16 is four times greater than four.

4 counted four times is the same as 16;
4 times 4 equals 16; 4 over and 4 up is 16.

0 × 0	0 × 1	0 × 2	0 × 3	0 × 4	0 × 5	0 × 6	0 × 7	0 × 8	0 × 9	0 × 10
1 × 0	1 × 1	1 × 2	1 × 3	1 × 4	1 × 5	1 × 6	1 × 7	1 × 8	1 × 9	1 × 10
2 × 0	2 × 1	2 × 2	2 × 3	2 × 4	2 × 5	2 × 6	2 × 7	2 × 8	2 × 9	2 × 10
3 × 0	3 × 1	3 × 2	3 × 3	3 × 4	3 × 5	3 × 6	3 × 7	3 × 8	3 × 9	3 × 10
4 × 0	4 × 1	4 × 2	4 × 3	4 × 4	4 × 5	4 × 6	4 × 7	4 × 8	4 × 9	4 × 10
5 × 0	5 × 1	5 × 2	5 × 3	5 × 4	5 × 5	5 × 6	5 × 7	5 × 8	5 × 9	5 × 10
6 × 0	6 × 1	6 × 2	6 × 3	6 × 4	6 × 5	6 × 6	6 × 7	6 × 8	6 × 9	6 × 10
7 × 0	7 × 1	7 × 2	7 × 3	7 × 4	7 × 5	7 × 6	7 × 7	7 × 8	7 × 9	7 × 10
8 × 0	8 × 1	8 × 2	8 × 3	8 × 4	8 × 5	8 × 6	8 × 7	8 × 8	8 × 9	8 × 10
9 × 0	9 × 1	9 × 2	9 × 3	9 × 4	9 × 5	9 × 6	9 × 7	9 × 8	9 × 9	9 × 10
10 × 0	10 × 1	10 × 2	10 × 3	10 × 4	10 × 5	10 × 6	10 × 7	10 × 8	10 × 9	10 × 10

LESSON 17

Skip Count by 7, Multiples of 10

Before multiplying by seven, learning to skip count by sevens will give us our foundation. Use the same techniques to introduce these facts that you have used before. Don't forget, if the student learns them by singing songs, make sure he or she can also say them without the aid of music.

Example 1

Skip count and write the numbers in the boxes with the lines. Whisper as you count the squares and then say the last number out loud as you write it. Rewrite the numbers in the blanks below the rectangle.

						7
						14
						21
						28
						35
						42
						49
						56

Notice that this rectangle illustrates a specific problem. It is $7 \times 8 = 56$.

7 , 14 , 21 , 28 , 35 , 42 , 49 , 56

Multiplication by Multiples of 10

When multiplying by multiples of 10 such as 20, 30, or 40, I like to multiply the factor times the digit in the tens place and then add the zero from the units place. To do this, I use a mitten that represents the teacher's hand to cover the zero when initially multiplying. Then I remove the mitten and add the zero.

Example 2
Multiply 20 times 4.

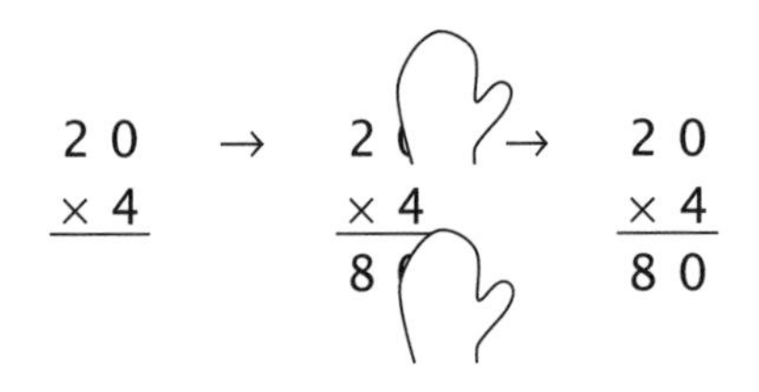

Example 3
Multiply 30 times 2.

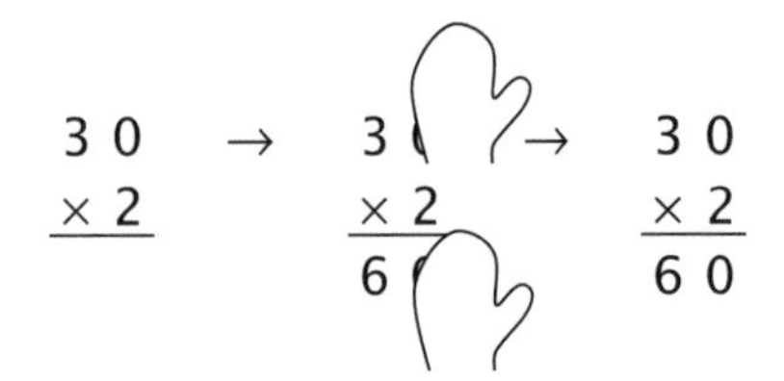

LESSON 18

Multiply by 7 and by Multiples of 100

There is one clever way to learn 7×8. Write the numbers in sequence: 5–6–7–8. Do you see the multiplication problem? Supply an equal sign and a times symbol, and you have $56 = 7 \times 8$. The only facts in this lesson that you have not been exposed to before are 7×7 and 7×8. Review all the seven facts that you have previously learned and focus on the two new ones before moving to the next lesson.

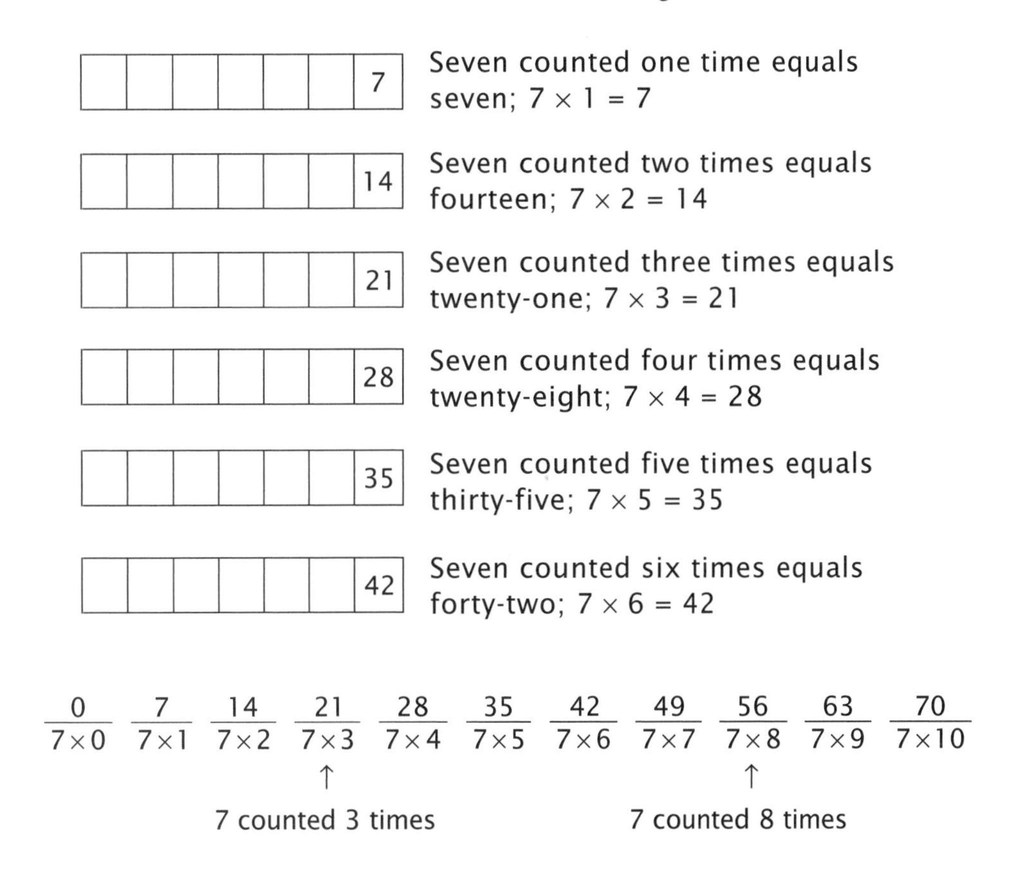

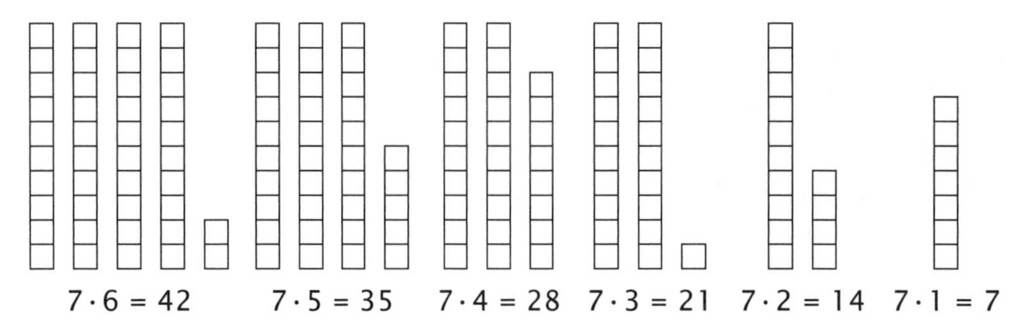

Another way to show this is on a number chart. Here we circle all of the seven facts, or multiples of seven.

0	1	2	3	4	5	6	**7**	8	9
10	11	12	13	**14**	15	16	17	18	19
20	**21**	22	23	24	25	26	27	**28**	29
30	31	32	33	34	**35**	36	37	38	39
40	41	**42**	43	44	45	46	47	48	**49**
50	51	52	53	54	55	**56**	57	58	59
60	61	62	**63**	64	65	66	67	68	69
70									

Each seven fact can be built in the shape of a rectangle. Whenever illustrating with the blocks, also write the problem and say it as you build. The drawing also illustrates the fact that 42 is seven times greater than six.

7 counted six times is the same as 42;
7 times 6 equals 42; 7 over and 6 up is 42.

0 × 0	0 × 1	0 × 2	0 × 3	0 × 4	0 × 5	0 × 6	0 × 7	0 × 8	0 × 9	0 × 10
1 × 0	1 × 1	1 × 2	1 × 3	1 × 4	1 × 5	1 × 6	1 × 7	1 × 8	1 × 9	1 × 10
2 × 0	2 × 1	2 × 2	2 × 3	2 × 4	2 × 5	2 × 6	2 × 7	2 × 8	2 × 9	2 × 10
3 × 0	3 × 1	3 × 2	3 × 3	3 × 4	3 × 5	3 × 6	3 × 7	3 × 8	3 × 9	3 × 10
4 × 0	4 × 1	4 × 2	4 × 3	4 × 4	4 × 5	4 × 6	4 × 7	4 × 8	4 × 9	4 × 10
5 × 0	5 × 1	5 × 2	5 × 3	5 × 4	5 × 5	5 × 6	5 × 7	5 × 8	5 × 9	5 × 10
6 × 0	6 × 1	6 × 2	6 × 3	6 × 4	6 × 5	6 × 6	6 × 7	6 × 8	6 × 9	6 × 10
7 × 0	7 × 1	7 × 2	7 × 3	7 × 4	7 × 5	7 × 6	7 × 7	7 × 8	7 × 9	7 × 10
8 × 0	8 × 1	8 × 2	8 × 3	8 × 4	8 × 5	8 × 6	8 × 7	8 × 8	8 × 9	8 × 10
9 × 0	9 × 1	9 × 2	9 × 3	9 × 4	9 × 5	9 × 6	9 × 7	9 × 8	9 × 9	9 × 10
10 × 0	10 × 1	10 × 2	10 × 3	10 × 4	10 × 5	10 × 6	10 × 7	10 × 8	10 × 9	10 × 10

Days in a Week

The most common application of multiplying by seven is the number of days in a week.

Example 1

How many days are in six weeks?

$7 \times 6 = 42$ days

Example 2

How many days are in seven weeks?

$7 \times 7 = 49$ days

Multiplication by Multiples of 100

When multiplying by multiples of 100, I like to multiply the factor times the digit in the hundreds place and then add the two zeros from the tens and units places. To do this, I use a mitten that represents the teacher's hand to cover the zeros when initially multiplying. Then I remove the mitten and add the zeros.

Example 3

Multiply 200 times 4.

$$\begin{array}{r} 200 \\ \times\ \ 4 \\ \hline \end{array} \rightarrow \begin{array}{r} 2\ \ \ \ \\ \times\ \ 4 \\ \hline 8\ \ \ \ \end{array} \rightarrow \begin{array}{r} 200 \\ \times\ \ 4 \\ \hline 800 \end{array}$$

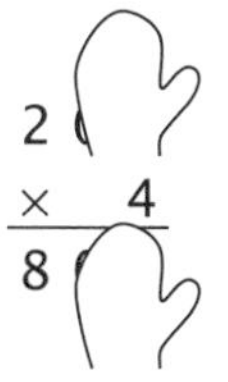

Example 4
Multiply 300 times 2.

$$\begin{array}{r} 300 \\ \times \quad 2 \\ \hline \end{array} \rightarrow \begin{array}{r} 3 \\ \times \quad 2 \\ \hline 6 \end{array} \rightarrow \begin{array}{r} 300 \\ \times \quad 2 \\ \hline 600 \end{array}$$

Example 5
Multiply 100 times 8.

$$\begin{array}{r} 100 \\ \times \quad 8 \\ \hline \end{array} \rightarrow \begin{array}{r} 1 \\ \times \quad 8 \\ \hline 8 \end{array} \rightarrow \begin{array}{r} 100 \\ \times \quad 8 \\ \hline 800 \end{array}$$

LESSON 19

Skip Count by 8, 8 Pints = 1 Gallon

Before we multiply by eight, skip counting by eight will get us started. Use the same techniques that you have used before in order to introduce these facts. Don't forget, if the student learns them by singing songs, make sure he or she can also say them without the aid of music.

Real-life examples for the eights are sides on a stop sign, pints in a gallon, legs on a spider, and arms on an octopus. ("Octo" means eight. An ***octagon*** has eight sides, and an octopus has eight arms.)

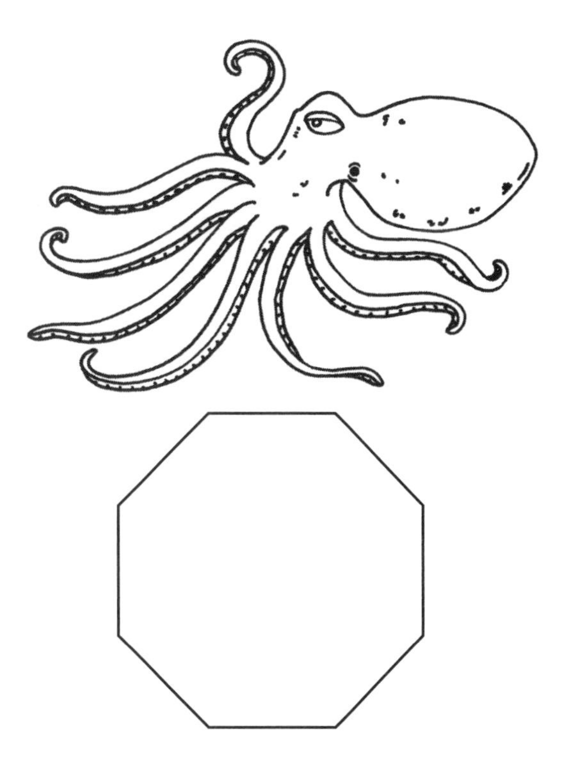

Example 1
Skip count and write the number in the boxes with the lines. Whisper as you count the squares and then say the last number out loud as you write it.

After filling in the rectangle, write the numbers in the spaces provided beneath the figure.

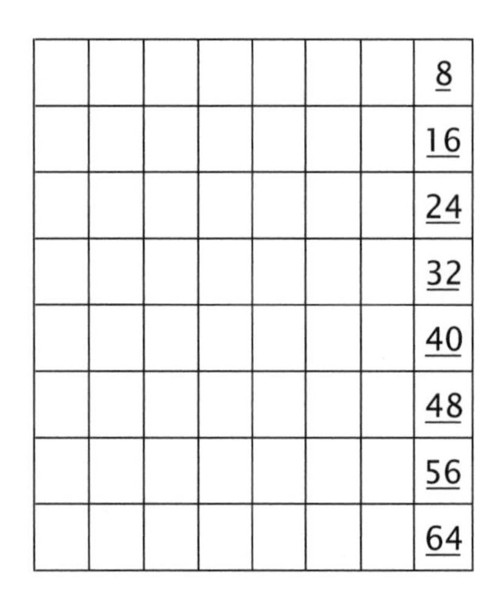

8 , 16 , 24 , 32 , 40 , 48 , 56 , 64

Notice that this rectangle illustrates a specific problem: $8 \times 8 = 64$

LESSON 20

Multiply by 8

There is only one fact you haven't studied yet, and that is 8 x 8. Review all the eight facts that you have previously learned and focus on this one before moving to the next lesson.

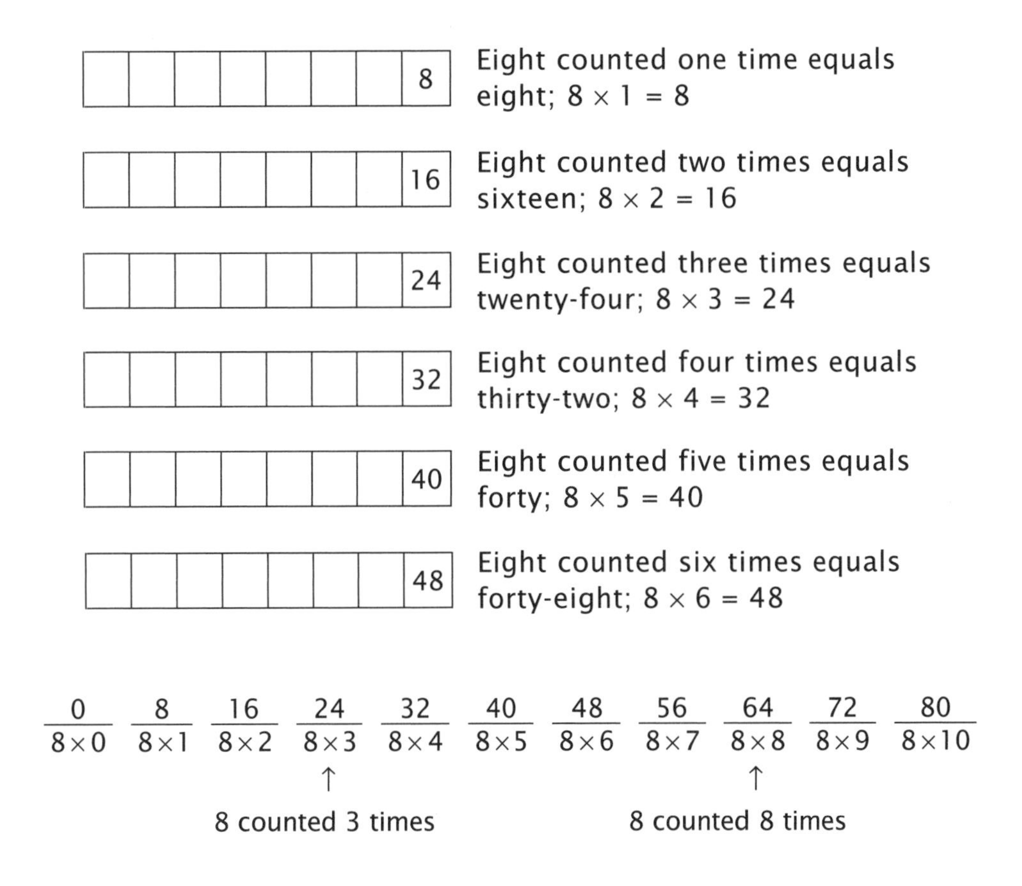

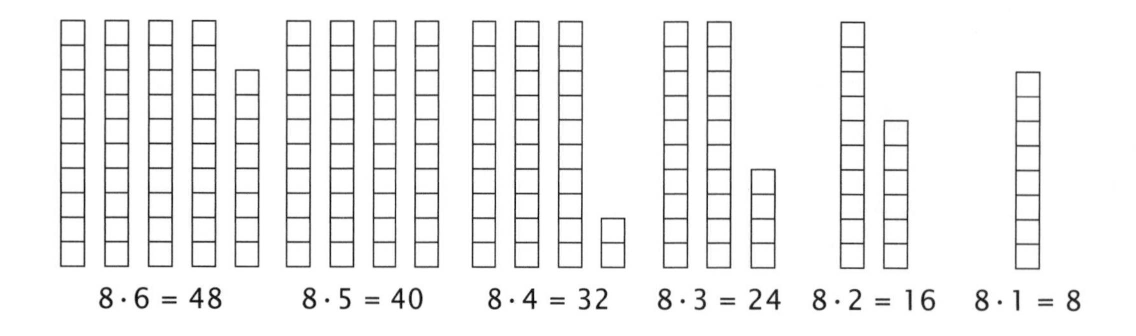

Another way to show this is on a number chart. Here we circle all of the eight facts, or multiples of eight.

0	1	2	3	4	5	6	7	8	9
10	11	12	13	14	15	16	17	18	19
20	21	22	23	24	25	26	27	28	29
30	31	32	33	34	35	36	37	38	39
40	41	42	43	44	45	46	47	48	49
50	51	52	53	54	55	56	57	58	59
60	61	62	63	64	65	66	67	68	69
70	71	72	73	74	75	76	77	78	79
80									

When illustrating with the blocks, remember to write the problem and say it as you build. The drawing also shows that 64 is eight times greater than eight.

8 counted eight times is the same as 64; 8 times 8 equals 64; 8 over and 8 up is 64.

0 × 0	0 × 1	0 × 2	0 × 3	0 × 4	0 × 5	0 × 6	0 × 7	0 × 8	0 × 9	0 × 10
1 × 0	1 × 1	1 × 2	1 × 3	1 × 4	1 × 5	1 × 6	1 × 7	1 × 8	1 × 9	1 × 10
2 × 0	2 × 1	2 × 2	2 × 3	2 × 4	2 × 5	2 × 6	2 × 7	2 × 8	2 × 9	2 × 10
3 × 0	3 × 1	3 × 2	3 × 3	3 × 4	3 × 5	3 × 6	3 × 7	3 × 8	3 × 9	3 × 10
4 × 0	4 × 1	4 × 2	4 × 3	4 × 4	4 × 5	4 × 6	4 × 7	4 × 8	4 × 9	4 × 10
5 × 0	5 × 1	5 × 2	5 × 3	5 × 4	5 × 5	5 × 6	5 × 7	5 × 8	5 × 9	5 × 10
6 × 0	6 × 1	6 × 2	6 × 3	6 × 4	6 × 5	6 × 6	6 × 7	6 × 8	6 × 9	6 × 10
7 × 0	7 × 1	7 × 2	7 × 3	7 × 4	7 × 5	7 × 6	7 × 7	7 × 8	7 × 9	7 × 10
8 × 0	8 × 1	8 × 2	8 × 3	8 × 4	8 × 5	8 × 6	8 × 7	8 × 8	8 × 9	8 × 10
9 × 0	9 × 1	9 × 2	9 × 3	9 × 4	9 × 5	9 × 6	9 × 7	9 × 8	9 × 9	9 × 10
10 × 0	10 × 1	10 × 2	10 × 3	10 × 4	10 × 5	10 × 6	10 × 7	10 × 8	10 × 9	10 × 10

A stop sign is shaped like an octagon. It has eight sides.

Example 1
How many sides are on eight stop signs?

8 × 8 = 64 sides

Example 2
How many sides are on seven stop signs?

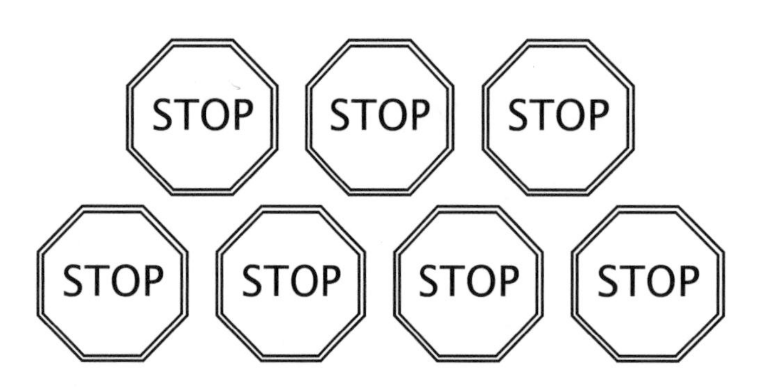

$8 \times 7 = 56$ sides

More on Solving for the Unknown and Division

Division will be taught in detail in *Delta*, the next book in the Math-U-See sequence. Most students are already familiar with using the word "divide" in everyday life. For example, they may discuss dividing a treat into two parts or dividing a group of items evenly among two or three people. Using this language informally is important for understanding the concept of division.

We strongly believe that it is preferable for students to master multiplication before learning division. Because of the practice provided at this level in solving for the unknown, students usually learn the division facts easily in *Delta*. However, some students may see division in testing situations before reaching *Delta*. If necessary, introduce the division symbol and show the student how division is related to solving for the unknown.

Solving for the unknown problems may be written several ways. For example, 3G = 18 means the same thing as $3 \times \_\_ = 18$. The student thinks, "Three times what will give me 18?" Of course, the answer is six.

Division turns this around and starts with the total, 18. If the 18 objects are divided into three groups, how many will be in each group? Write the problem as $18 \div 3 = \_\_\_\_$. The answer is six because if we have three groups of six, we can multiply 3×6 to get back to 18.

Use blocks or other objects and the idea of equal shares to help students understand the conceptual basis of division. Relate division to multiplication by multiplying the number of groups by the number of objects in each group to get back the original amount.

LESSON 21

Multiple-Digit Multiplication

Place-Value Notation, Distributive Property

Place-value notation is a way to write out numbers by showing the places separately. It is related to the way we taught place value with the blocks. For example, 123 can be written as 100 + 20 + 3. This notation reinforces place value. Because it represents the blocks so well, we will use place-value notation to work each example, as well as standard notation and the manipulatives (whenever possible).

Remember that the ***factors*** are the outside dimensions of a rectangle and the ***product*** is the area of a rectangle. In Example 1 on the next page, notice that the up factor is on the line with the multiplication symbol and the over factor is on the top line. Switching these factors will still produce the same answer, but it won't correspond to the picture.

Another skill presented here for the first time is the ***Distributive Property of Multiplication over Addition.*** In Example 1, when we multiply 23 by 2, we are multiplying the 2 times the 20 and 2 times the 3. The 2 is "distributed" between both components, or place values, present in 23. Verbalize this as "Twenty-three counted two times is the same as twenty counted two times plus three counted two times." Figure 1 shows 23 multiplied by 2 horizontally. While you may multiply this way, in this book we will usually multiply vertically, as in Example 1 on the next page.

Figure 1
Distributive Property of Multiplication over Addition

$$\begin{aligned} 2(23) &= 2(20+3) \\ &= 2\cdot 20 + 3\cdot 2 \\ &= 40 + 6 = 46 \end{aligned}$$

Example 1

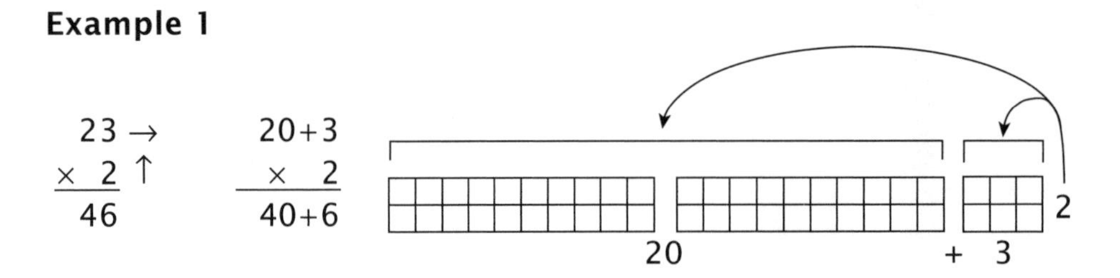

Notice that in Figure 1 and Example 1, we first multiply the 2 times the units (3) and then times the tens (20).

Example 2

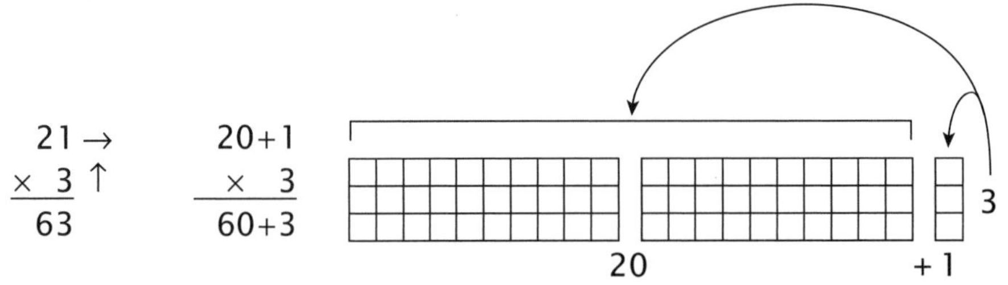

The next examples are too large to show with the blocks, but we can use the same techniques as in the first two examples. On the right side of each example, the parentheses in the second line show us that we multiply each part of the problem separately and then add the answers.

Example 3

$$\begin{array}{r} 231 \rightarrow \\ \times \quad 3 \uparrow \\ \hline 693 \end{array} \qquad \begin{array}{r} 200+30+1 \\ \times \qquad\quad 3 \\ \hline 600+90+3 \end{array} \qquad \begin{aligned} 3(231) &= 3(200+30+1) \\ &= (3 \cdot 200)+(3 \cdot 30)+(3 \cdot 1) \\ &= 600+90+3 = 693 \end{aligned}$$

Example 4

$$\begin{array}{r} 412 \rightarrow \\ \times \quad 2 \uparrow \\ \hline 824 \end{array} \qquad \begin{array}{r} 400+10+2 \\ \times \qquad\quad 2 \\ \hline 800+20+4 \end{array} \qquad \begin{aligned} 2(412) &= 2(400+10+2) \\ &= (2 \cdot 400)+(2 \cdot 10)+(2 \cdot 2) \\ &= 800+20+4 = 824 \end{aligned}$$

When solving word problems, the student can turn a piece of notebook paper sideways and use the lines to keep the digits lined up. This is especially useful when doing problems with greater numbers.

LESSON 22

Rounding to 10, 100, and 1,000

Estimation

Most of this lesson should be review, as we covered estimation in the previous book. If rounding is new, however, take the time you need to understand it.

Rounding is used to estimate an answer before we multiply. When you round a number to the nearest multiple of 10, there will be a number between one and nine in the tens place but only a zero in the units place.

Let's round 38 to the nearest 10 as an example. The first step is to find the two multiples of 10 that are nearest to 38. The lesser one is 30, and the greater one is 40 because thirty-eight is between 30 and 40. If the student has trouble finding these numbers, begin by placing your finger over the 8 in the units place so that all you have is a 3 in the tens place, which is 30. Then add one more to the tens to find the 40. I often write the numbers 30 and 40 above the number 38 on both sides, as in Figure 1.

Figure 1

```
30   40
  38
```

The next step is to find out whether 38 is closer to 30 or 40. Let's go through all the numbers, as shown in Figure 2. It is obvious that 31, 32, 33, and 34 are closer to 30 and that 36, 37, 38, and 39 are closer to 40, but 35 is a special case. It is just as close to 30 as it is to 40. There are different rules that can be used with numbers ending in 5, but the rule we will use here is to round numbers with a 5 in the units place up to the nearest ten. When rounding to tens, look at the units place. If the units are 0, 1, 2, 3, or 4, the digit in the tens place remains unchanged. If the units are 5, 6, 7, 8, or 9, the digit in the tens place increases by one.

Figure 2

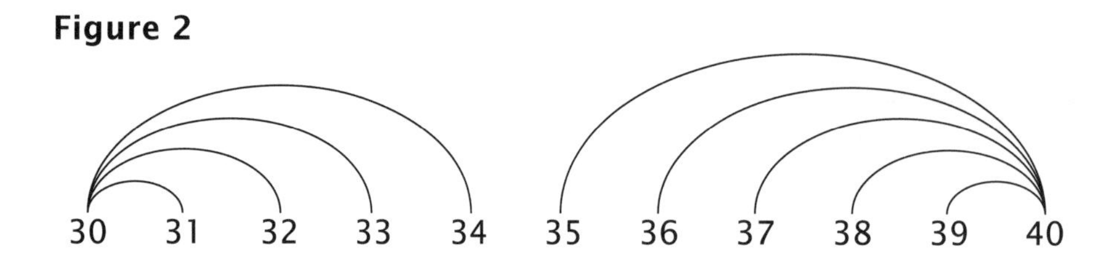

Another strategy I use is to put 0, 1, 2, 3, and 4 inside a circle to represent zero because, if these digits are in the units place, they add nothing to the tens place. They are rounded to the lesser number (30 in the example). Then I put 5, 6, 7, 8, and 9 inside a thin rectangle to represent one because, if these digits are in the units place, they add one to the tens place and are rounded to the greater number (40 in the example).

Figure 3

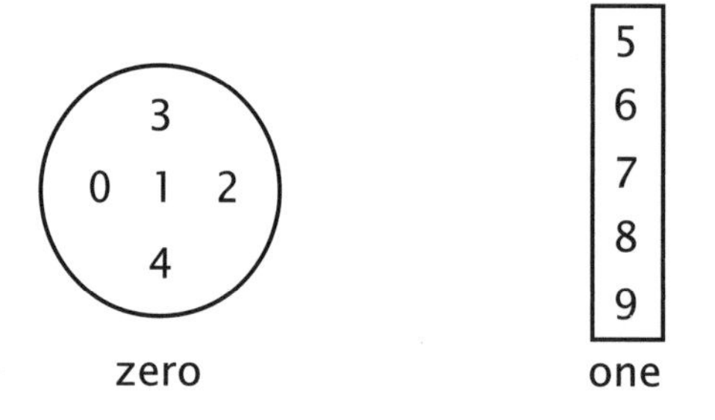

Example 1
Round 43 to the nearest tens place.

40 50 43	1. Find the multiples of 10 nearest to 43.
40 ← 50 43	2. We know that 3 goes to the lesser number, which is 40.
(40) ← 50 43	3. Recall that 3 is in the circle, or 0, so nothing is added to the lesser number, 40.

When rounding to hundreds, look only at the digit in the tens place to determine whether to stay the same or increase by one. The same rules apply to hundreds as to tens. If the digit in the tens place is 0, 1, 2, 3, or 4, the digit in the hundreds place remains unchanged. If the digit in the tens place is a 5, 6, 7, 8, or 9, then the digit in the hundreds place increases by 1.

When we look at rounding a number like 635, we can see one of the reasons why we recommend rounding 5s up instead of down. Although 650 is exactly halfway between 600 and 700, all the other 650s (651, 652, and so on) are closer to 700. Rounding 5s upward makes sense because if there is another non-zero digit, the number will be closer to the greater number.

Example 2
Round 547 to the nearest hundreds place.

500 600
547

1. Find the multiples of 100 nearest to 547.

500 ↰ 600
547

2. We know that 4 goes to the lesser number, 500.

(500) ↰ 600
547

3. Recall that 4 is in the circle, or 0, so nothing is added to the lesser number, 500.

When rounding to thousands, consider only the digit immediately to the right of the thousands place (the hundreds) to determine whether to stay the same or increase by one.

Example 3
Round 8,719 to the nearest thousands place.

8,000 9,000
8,719

1. Find the nearest multiples of 1,000.

8,000 ↱ 9,000
8,719

2. We know that 7 goes to the greater number, 9,000.

8,000 ↱ (9,000)
8,719

3. Recall that 7 is in the rectangle, or 1, so 1 is added to the 8, and the answer is 9,000.

Estimation

Now that we know how to round numbers, we can apply this skill to find the approximate answer for a multiplication problem. In Example 4, we are going to round only the greater factor. Later, when both factors have multiple digits, we will round both of them.

Example 4

Estimate the answer.

$$\begin{array}{r} 43 \\ \times\ 2 \\ \hline \end{array} \qquad \begin{array}{r} (40) \\ \times\ \ 2 \\ \hline (80) \end{array}$$

Round the number and put it in the parentheses.

Multiply to find the approximation.

Example 5

Estimate the answer.

$$\begin{array}{r} 281 \\ \times\ 3 \\ \hline \end{array} \qquad \begin{array}{r} (300) \\ \times\ \ \ 3 \\ \hline (900) \end{array}$$

Round the number to the hundreds place and put it in the parentheses.

Multiply to find the approximation.

Example 6

Estimate the answer.

$$\begin{array}{r} 419 \\ \times\ 2 \\ \hline \end{array} \qquad \begin{array}{r} (400) \\ \times\ \ \ 2 \\ \hline (800) \end{array}$$

Round the number to the hundreds place and put it in the parentheses.

Multiply to find the approximation.

Double-Digit Times Double-Digit

Multiplication by 11

When multiplying two double-digit numbers, you actually have four multiplication problems. Notice the four different rectangles. Begin with the units times both of the top (over) factors and then multiply the tens times both of the top factors. The four smaller multiplication problems, or ***partial products***, are 2 × 3, 2 × 10 (not 2 × 1, but two times one ten), 10 × 3, and 10 × 10. Study the picture and the written portion to understand this. When using the units, tens, and hundreds blocks, the partial products are very clear.

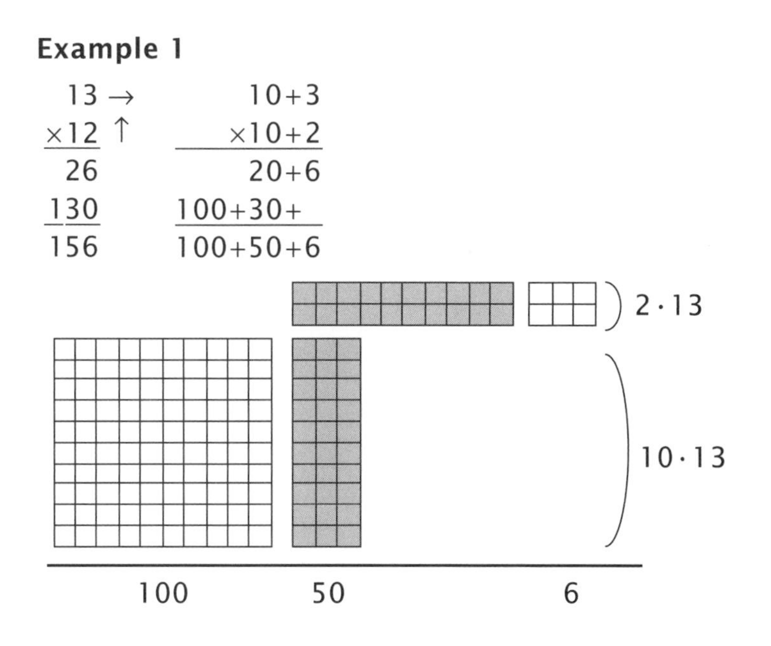

Remember that each partial product is also the area of one of the small rectangles. Area is sometimes referred to as ***additive*** because, if we find the areas of the separate rectangles and add them together, we will have the area of the original larger rectangle.

We see in Example 1 that the problem 12×13 can also be broken down into two problems, 10×13 and 2×13. Double-digit multiplication combines these two problems into one. You can see this in the figure on the previous page, with the factors to the right of the drawing.

$$\begin{array}{r} 13 \rightarrow \\ \underline{\times 10} \uparrow \\ 130 \end{array} \qquad \begin{array}{r} 13 \rightarrow \\ \underline{\times\ \ 2} \uparrow \\ 26 \end{array}$$

Have you ever wondered why you shift the second line of a multiplication problem? It is because of place value. After building the rectangle, ask the student, "Is every value is in its own place?" In the written problem, we wrote zeros to fill place values. However, if we understand place value, we don't need to write all the zeros but just put everything in its place. The reason for the shift to the left is to put every value in its own place so that we can add the units to the units, the tens to the tens, and the hundreds to the hundreds.

Example 2

$$\begin{array}{r} 14 \\ \underline{\times 12} \\ 28 \\ \underline{140} \\ 168 \end{array} \qquad \begin{array}{r} 10+4 \\ \underline{\times 10+2} \\ 20+8 \\ \underline{100+40} \\ 100+60+8 \end{array}$$

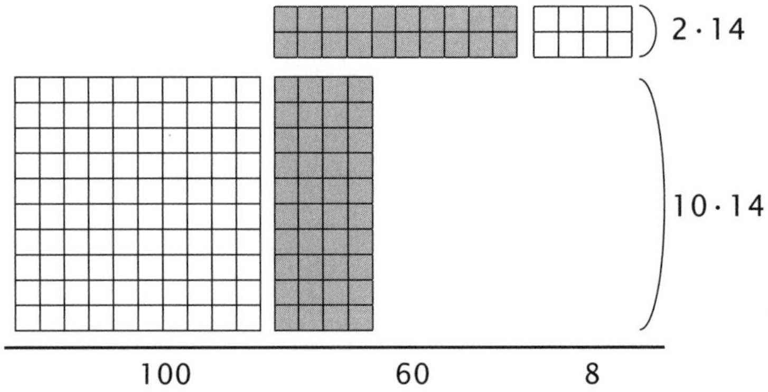

$2 \cdot 14$

$10 \cdot 14$

100 60 8

Example 3

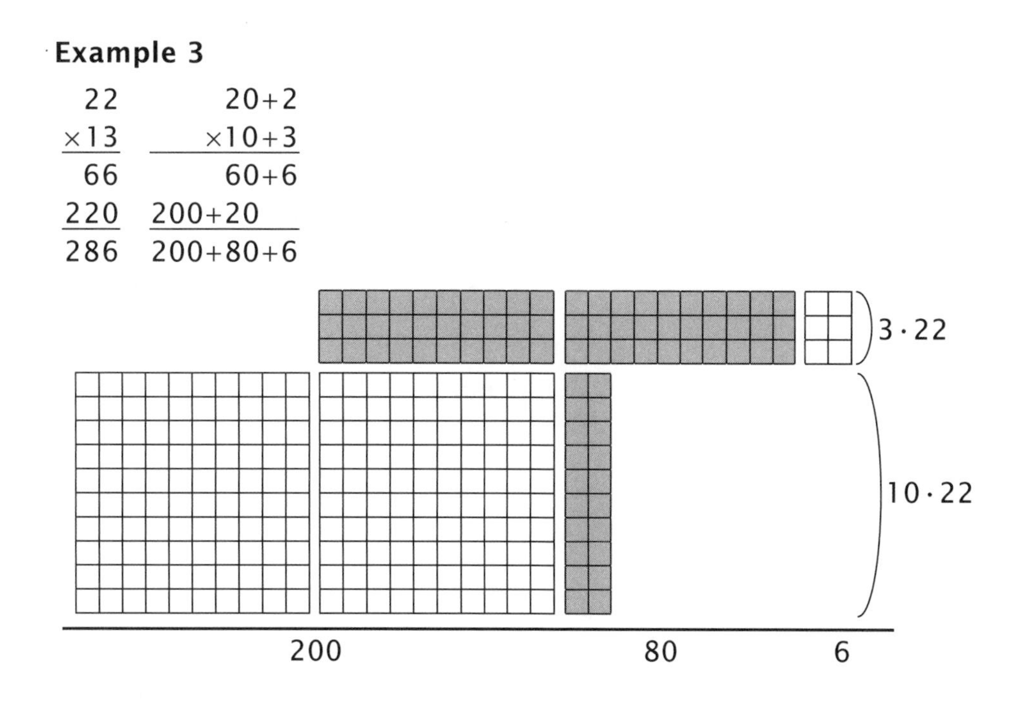

Example 4 is an example of a three-digit number times a two-digit number. Notice that it is a combination of what we have been doing. Notice also how it is two problems being solved together.

Example 4

212		212		212		200+10+2
×10	+	× 4	=	×14	or	× 10+4
2,120		848		848		800+40+8
				2,120		2,000+100+20
				2,968		2,000+900+60+8

Multiplication by 11

When you multiply by 11, there is an interesting phenomenon that occurs. I call it "split 'em and add 'em." This stategy only works if the number being multiplied by 11 has only two digits and those digits don't add up to more than 9.

Examples 5 and 6 on the next page meet these criteria. The pattern is to "split 'em" (the digits in 23) to 2 _ 3, and then "add 'em," 2 + 3 = 5, for the middle term. You get 253. Looking at the examples, do you see where this originates? When you multiply by 11, the digits in the partial product are the same. This is not critical to learn, but it's an interesting observation of number patterns.

Example 5

$$\begin{array}{r} 23 \\ \times 11 \\ \hline 23 \\ 23 \\ \hline 253 \end{array}$$

same → 23
→ 23

↗↑↖

split 'em↑ split 'em

add 'em

Example 6

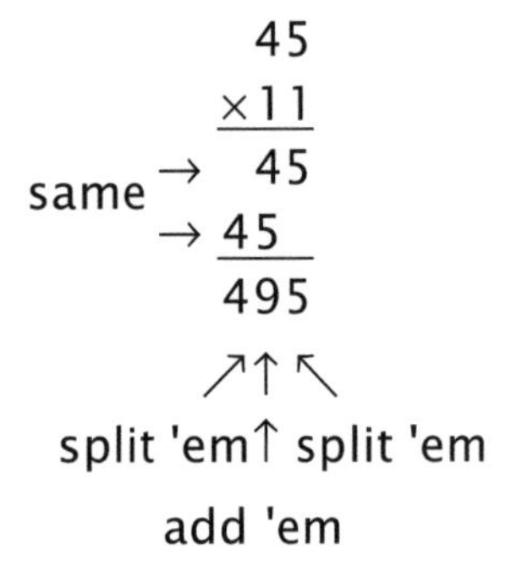

LESSON 24

Double-Digit with Regrouping

Associative Property; Mental Math

The way that I present this topic is not the traditional approach, but it has distinct advantages that are revealed by the manipulatives. Consider this method as a possible alternative way of working a multiplication problem with regrouping.

First, look at a single-digit times a double-digit problem with regrouping.

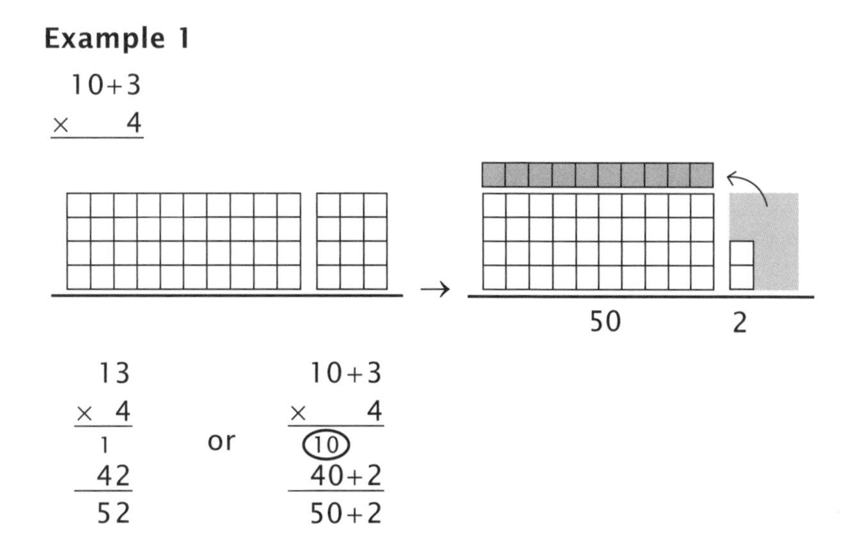

Multiplying gives 12 in the units place, so we need to regroup. The arrow and the shading show the 10 moving to the tens place. Instead of writing the 1 above the 1 in the tens place, put it in the tens place below the line. Then add as usual.

Example 2

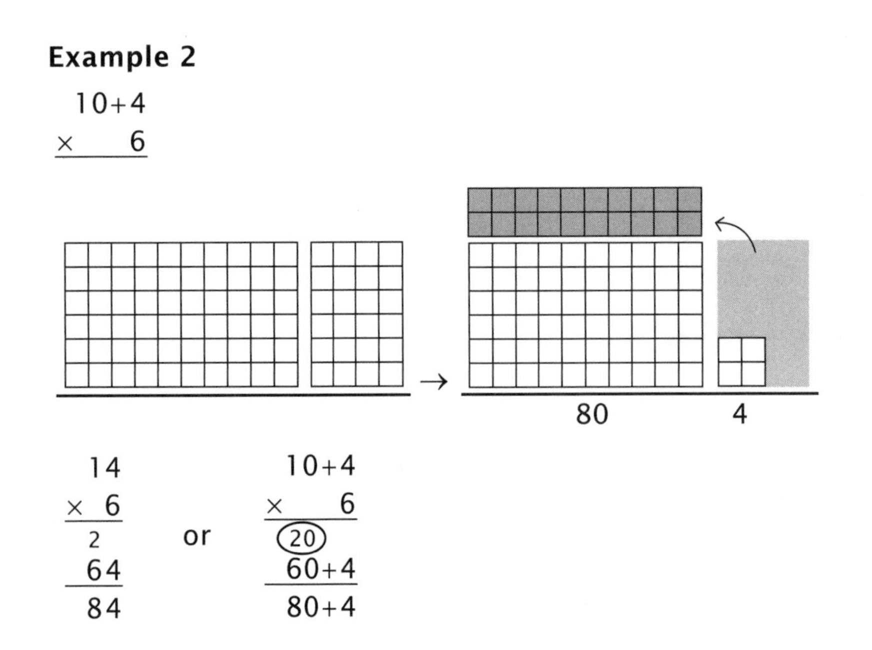

Multiplying gives 24 in the units place, so we need to regroup. Instead of writing the 2 above the 2 in the tens place, put it in the tens place below the line. Then add the numbers in each place.

Example 3

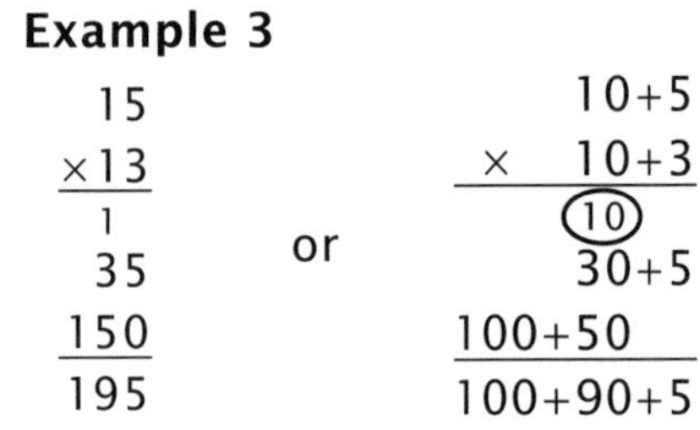

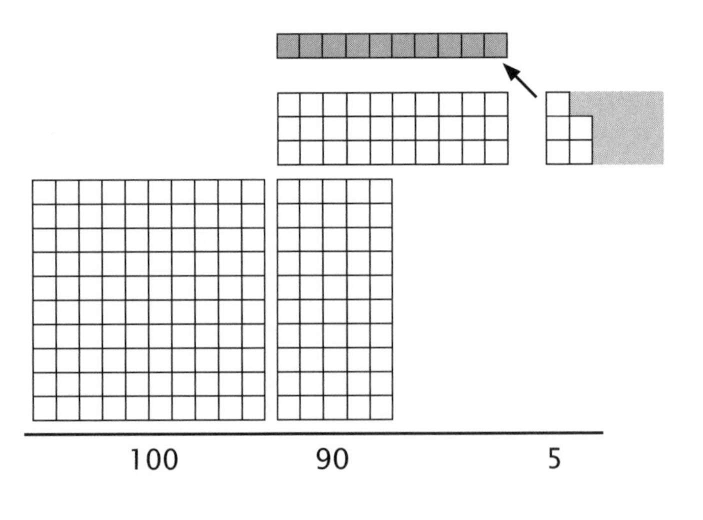

After multiplying, you have 15 in the units place and need to regroup. Instead of writing the 1 above the 1 in the tens place, put it in the tens place below the line. Then add as usual.

To help keep the place values properly aligned, here are two tips. First, either use graph paper or lined paper turned sideways to keep the digits in the proper place. Second, when you add, you can circle any digits that are regrouped. This helps remove potential confusion.

Example 4

24		20+4
×13		× 10+3
11	or	(100) 10
62		60+2
240		200+40
312		300+10+2

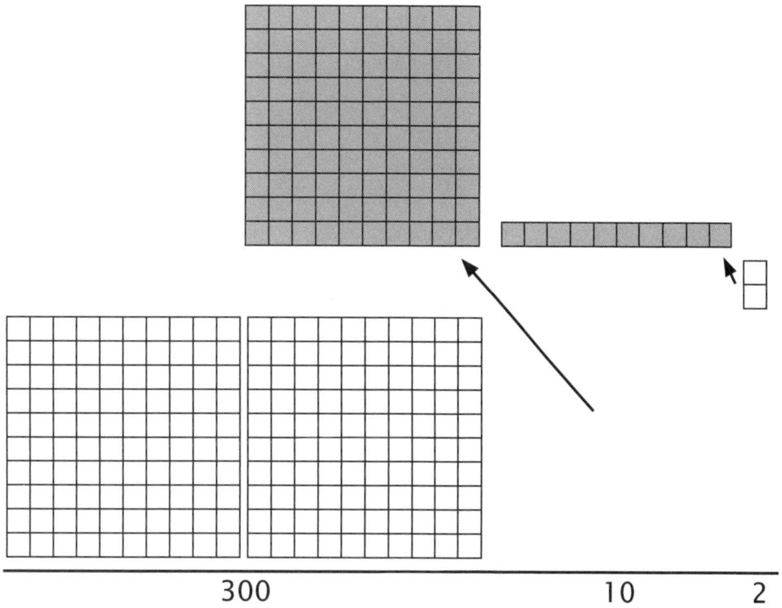

300 10 2

In Example 4, we regrouped the initial partial product, as shown by the arrow in the units place. When adding the tens after multiplying, we regroup 10 tens as 1 hundred. The hundred is shaded as the ten was in previous examples.

Using the traditional method, after you multiply the units, you place the 1 above the tens place, as shown in Example 5. Then you multiply 3 × 1 to get 3 and add the 1 to make it 4. There are two potential problems that this creates. First, many students never understand why you mix the operations and multiply 3 × 1 and then *add* the other 1. Second, in a problem with a lot of regrouping, you can write quite a few numbers above the top factor, and then the student adds the wrong numbers.

With my method, you do all the multiplying first and then do all the adding below the line. In some cases, it is not only clearer but quicker. Some older students who have already learned the traditional way won't like to switch, so show the method to them and make it optional. Both ways work.

Example 5

		traditional way
15	10+5	1
×13	× 10+3	15
1	10	×13
35	30+5	45
150	100+50	15
195	100+90+5	195

Here are more examples to study. You decide the method you prefer your student to use. These problems are too large to show with the blocks.

Example 6

		traditional way
37	30+7	2
×14	× 10+4	37
12	(100) (20)	×14
28	20+8	1
1	(100)	148
370	300+70	370
518	500+10+8	518

Example 7

		traditional way
28	20+8	15
×27	× 20+7	28
1 5	(100) (50)	×27
1 46	(100) 40+6	1
1	(100)	196
460	400+60	560
756	700+50+6	756

Associative Property

There are word problems in the student workbook that require three numbers to be multiplied together. The most straightforward way to solve these is to multiply the first two numbers and then multiply the product by the third number. Be sure the student understands that the numbers may be multiplied in any order with the same result. For example, a problem requires you to multiply 2, 3, and 8. Using parentheses to show what you do first, it looks like this:

$$\begin{array}{r} (2\times 3)\times 8 = ? \\ 6\times 8 = 48 \end{array} \quad \text{or} \quad \begin{array}{r} 2\times (3\times 8) = ? \\ 2\times 24 = 48 \end{array}$$

Mental Math

These problems can be used to keep the multiplication facts fresh in the memory and to develop mental math skills. Try a few at a time, saying the problem slowly so that the student comprehends. Lessons 27 and 30 have more of these.

Example 8

"Two times three, times one, equals what number?"

The student thinks, "$2 \times 3 = 6$, and $6 \times 1 = 6$." At first, go slowly enough for him to verbalize the intermediate step. As skills increase, the student should be able to say just the answer.

1. Two times three, times eight, equals what number? (48)
2. One times seven, times five, equals what number? (35)
3. Four times two, times nine, equals what number? (72)
4. Three times three, times three, equals what number? (27)
5. Five times one, times six, equals what number? (30)
6. Two times two, times eight, equals what number? (32)
7. Three times two, times two, equals what number? (12)
8. One times nine, times seven, equals what number? (63)
9. Seven times two, times zero, equals what number? (0)
10. Five times four, times one, equals what number? (20)

LESSON 25

Multiple-Digit Multiplication, Regrouping

This is not much different from the previous lesson except that now we are multiplying three digits times two digits with regrouping. In Example 1, on the next page, we see that the problem is still a combination of two problems, since the multiplier has two digits (17). The top factor, or the number being multiplied, is the multiplicand, and the answer is the product. The estimates are in the parentheses.

Also in this lesson, we are using ***ten thousands***. Look at Figure 1. Notice that each grouping of three digits contains a pattern. The first three digits record hundreds, tens, and units. The next three record hundreds (of thousands), tens (of thousands), and units (of thousands). A comma separates these groupings of digits to make them easier to read. I like to think of the first comma as representing the word "thousand" when reading the number. For instance, the number 18,243 is read as "eighteen thousand, two hundred forty-three."

Figure 1

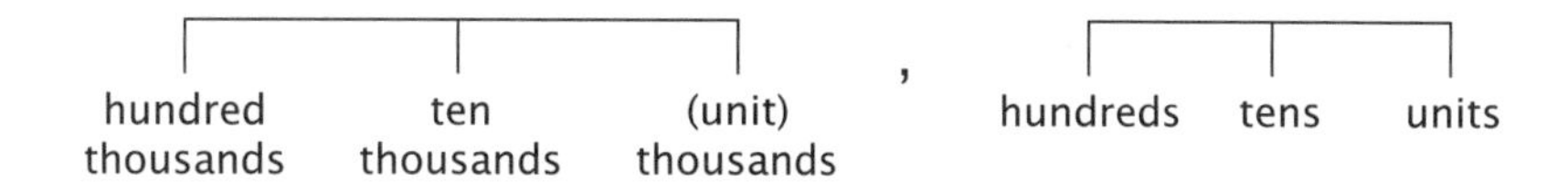

Example 1

$$\begin{array}{r} 245 \\ \times\ \ 7 \\ \hline {}^{1}23 \\ 1,485 \\ \hline 1,715 \end{array} \quad + \quad \begin{array}{r} 245 \\ \times 10 \\ \hline 2,450 \end{array} \quad = \quad \begin{array}{r} 245 \\ \times 17 \\ \hline 1\ {}^{1}23 \\ 1,485 \\ 2,450 \\ \hline 4,165 \end{array} \qquad \begin{array}{r} (200) \\ \times (20) \\ \hline (4,000) \end{array} \qquad \begin{array}{r} (2 \\ \times (2 \\ \hline (4,000) \end{array}$$

Example 2

$$\begin{array}{r} 163 \\ \times\ \ 9 \\ \hline 52 \\ {}^{1}947 \\ \hline 1,467 \end{array} \quad + \quad \begin{array}{r} 163 \\ \times 80 \\ \hline {}^{1}4\ 2 \\ 8,840 \\ \hline 13,040 \end{array} \quad = \quad \begin{array}{r} 163 \\ \times 89 \\ \hline {}^{1}52 \\ {}^{2}947 \\ 142\ \ \ \\ 8,840 \\ \hline 14,507 \end{array} \qquad \begin{array}{r} (200) \\ \times (90) \\ \hline (18,000) \end{array} \qquad \begin{array}{r} (2 \\ \times (9 \\ \hline (18,000) \end{array}$$

Example 3

$$\begin{array}{r} 412 \\ \times\ \ 7 \\ \hline 2\ \ 1\ \ \\ 874 \\ \hline 2,884 \end{array} \quad + \quad \begin{array}{r} 412 \\ \times 50 \\ \hline 1\ \ \ \\ 20,500 \\ \hline 20,600 \end{array} \quad = \quad \begin{array}{r} 412 \\ \times 57 \\ \hline 2\ \ 1\ \ \\ {}_{1}\ 874 \\ 1\ \ \ \\ 2{}^{1}0,500 \\ \hline 23,484 \end{array} \qquad \begin{array}{r} (400) \\ \times (60) \\ \hline (24,000) \end{array} \qquad \begin{array}{r} (4 \\ \times (6 \\ \hline (24,000) \end{array}$$

Example 4

$$\begin{array}{r} 208 \\ \times\ \ 3 \\ \hline 2\ \ \\ 604 \\ \hline 624 \end{array} \quad + \quad \begin{array}{r} 208 \\ \times 60 \\ \hline 4\ \ \ \ \\ 12,080 \\ \hline 12,480 \end{array} \quad = \quad \begin{array}{r} 208 \\ \times 63 \\ \hline 1\ 2\ \ \\ 604 \\ 1\ 4\ \ \ \ \ \ \\ 12,080 \\ \hline 13,104 \end{array} \qquad \begin{array}{r} (200) \\ \times (60) \\ \hline (12,000) \end{array} \qquad \begin{array}{r} (2 \\ \times (6 \\ \hline (12,000) \end{array}$$

LESSON 26

Finding Factors, 25¢ = 1 Quarter

Factoring is the opposite of multiplying. When you multiply, you are given two factors, and you find the product. When you factor, you are given the area, or product, and you find the factors. Let's work through the first example. Take six green unit blocks and see how many different rectangles you can make. In this case, you can build two rectangles, one by six and two by three. Therefore, the factors of six are 1, 2, 3, and 6.

Example 1
Find the factors of 6.

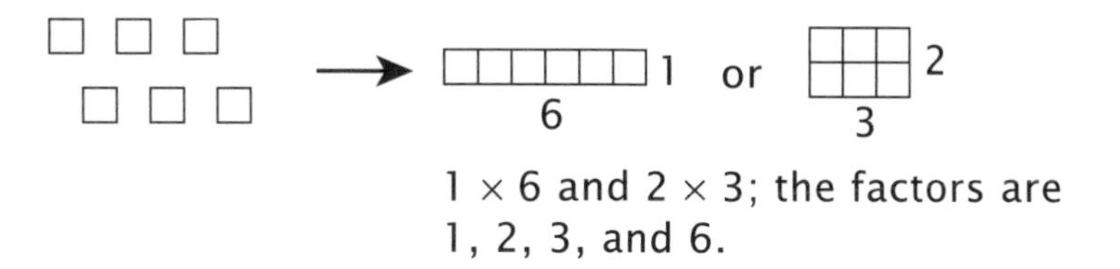

1 × 6 and 2 × 3; the factors are 1, 2, 3, and 6.

Example 2
Find the factors of 12.

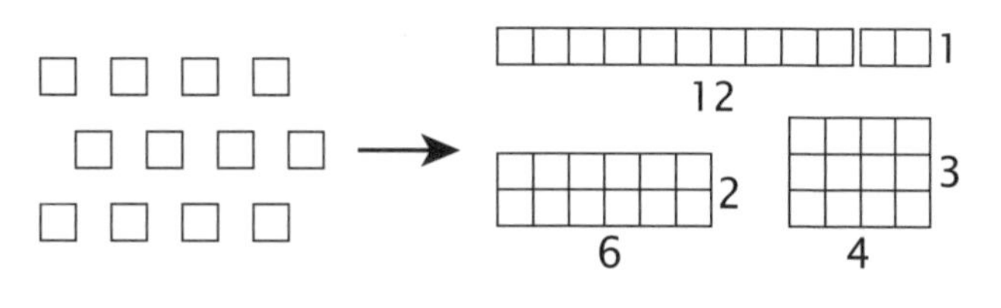

1 × 12, 2 × 6, and 3 × 4; the factors are 1, 2, 3, 4, 6, and 12.

Example 3
Find the factors of 15.

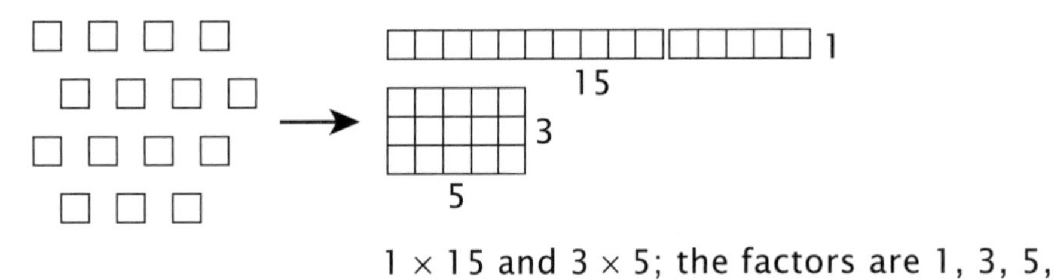

1 × 15 and 3 × 5; the factors are 1, 3, 5, and 15.

Example 4
Find the factors of 24.

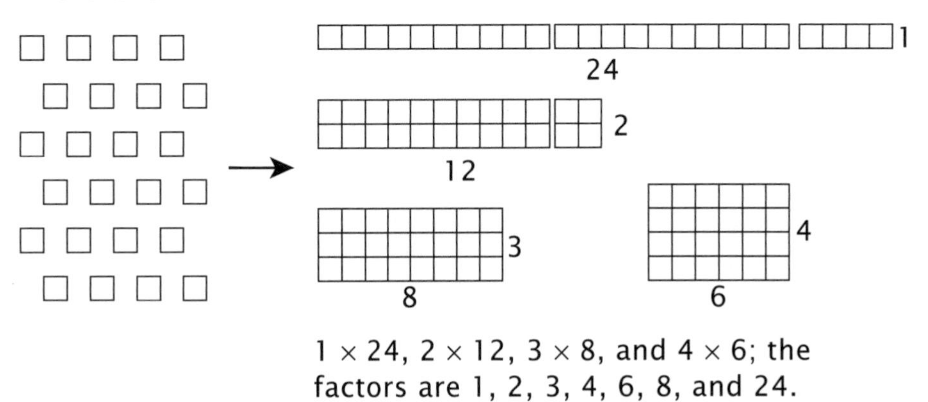

1 × 24, 2 × 12, 3 × 8, and 4 × 6; the factors are 1, 2, 3, 4, 6, 8, and 24.

When factoring, use what you know about the multiplication facts to help you find potential factors. If the number is even, then you know it will have a factor of two. If the digits add up to three or a multiple of three, then you know that three is a factor. If a number ends in zero, then we know that it is a multiple of 10. If a number ends in a five or a zero, then five is a factor, and if the digits add up to nine or a multiple of nine, then nine will also be a factor. Knowing your multiplication facts and recognizing patterns in multiplication is the best preparation for factoring.

Factoring Greater Numbers

Students should be able to find the factors of many numbers up to 100 by using their knowledge of multiplication and solving for the unknown. Some numbers will be easier to factor once division has been taught. However, all numbers up to 100 can be factored by using the blocks to build all possible rectangles.

25¢ = 1 Quarter of a Dollar

In lesson 16 we learned that there are four quarters in a dollar and that there are 25¢ in a quarter. Now that we know how to regroup while multiplying double-digit numbers, we can find out the number of cents in any number of quarters. See Examples 5, 6, 7, and 8.

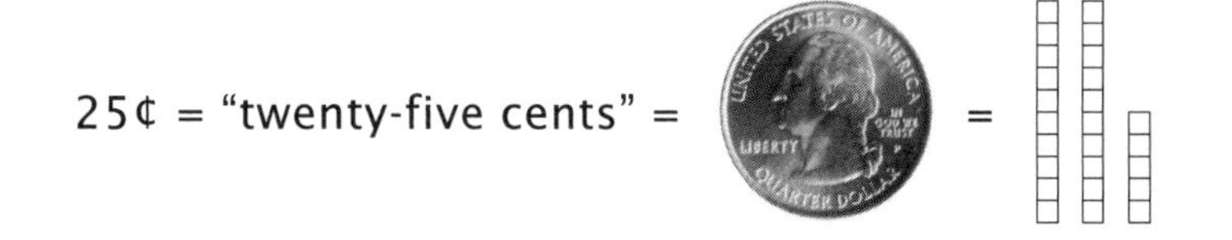

Example 5
How many cents are in three quarters?

$$\begin{array}{rr} 25 & 20+5 \\ \times\ 3 & \times\ 3 \\ \hline {}^{1} & {}^{1}\quad \\ 65 & 60+5 \\ \hline 75 & 70+5 \end{array}$$

Example 6
How many cents are in seven quarters?

$$\begin{array}{rr} 25 & 20+5 \\ \times\ 7 & \times\ 7 \\ \hline {}^{3} & {}^{30} \\ 145 & 100+40+5 \\ \hline 175 & 100+70+5 \end{array}$$

Example 7
How many cents are in 11 quarters? (Remember "split 'em and add 'em.")

$$\begin{array}{rr} 25 & 20+5 \\ \times 11 & \times 10+1 \\ \hline 25 & 20+5 \\ 25\ \ & 200+50\quad \\ \hline 275 & 200+70+5 \end{array}$$

Example 8
How many cents are in 25 quarters?

25	20+5
×25	×20+5
2	20
105	100+ 0+5
1	100
40	400+ 0
625	600+20+5

LESSON 27

Place Value Through Millions

16 Ounces = 1 Pound

Understanding place value helps you know where to place the digits in multiple-digit multiplication. For some, this will be review. The beginning value we call units. This is represented by the small green half-inch cube. The next greatest place value is the tens place, shown with the blue 10 bar. This is 10 times greater than a unit. The next value we come to, as we move to the left, is the hundreds, shown by the large red block. It is 10 times as large as a 10 bar. Notice that as you move to the left, each value is 10 times greater than the preceding value.

Figure 1

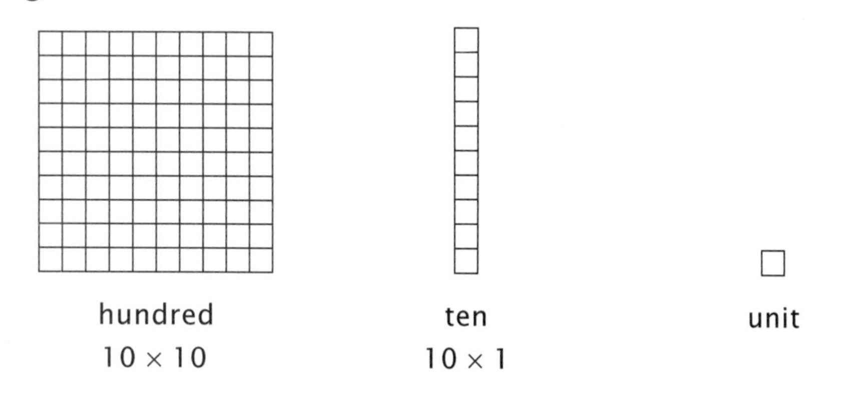

When you name a number such as 247, the 2 tells you how many hundreds, the 4 tells how many tens, and the 7 tells how many units. We read 247 as "two hundred for-ty seven." The letters "ty" mean 10, so four tens are forty. The 2, 4, and 7 are digits that tell us how many. The hundreds, tens, and units tell us what kind, or what value. Where the digit is written, or what ***place*** it occupies, tells us its ***value***. Notice that, as the values progress from right to left, they keep increasing by a factor of 10. That is because we are operating in the base 10 system, or the decimal system.

The next place value is the thousands place. It is 10 times 100. You can build 1,000 by stacking 10 hundred squares and making a cube. You can also show 1,000 by making a rectangle that is 10 by 100 out of the cube, as in Figure 2. You will see that I used a much smaller scale to be able to show this. Also in Figure 2, 10,000 is shown. If you continue to use rectangles, can you imagine what 1,000,000 would look like? It would be a rectangle 100 by 1,000. The factors are shown inside the rectangles.

Figure 2

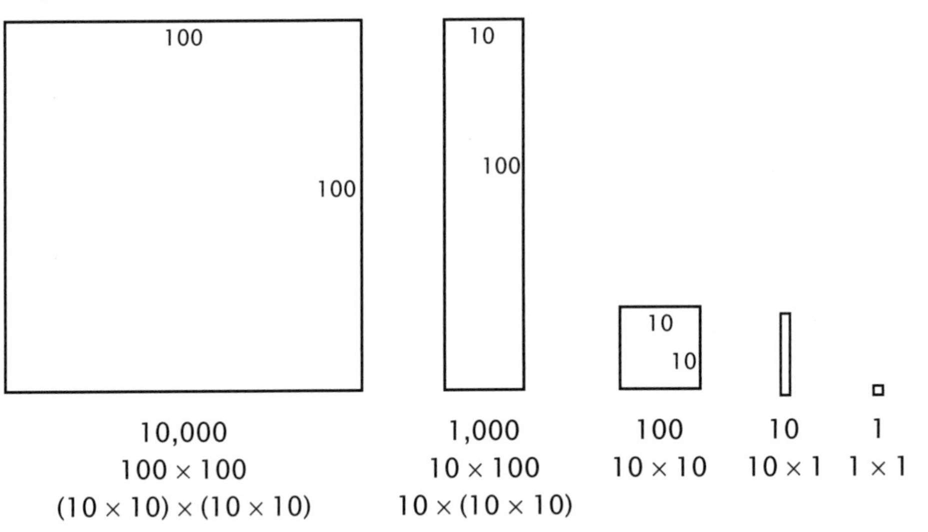

Notice the progression of multiplying by a factor of 10 as you move from right to left. Ten times 1,000 is 10,000, and 10 times 10,000 is 100,000. Ten times 100,000 is 1,000,000 (one million).

In Figure 3, do you see that there are three places within the thousands? The same is true with the millions. There are millions, 10 millions, and 100 millions. The commas separate the number into groups of three.

Figure 3

1 2 3	,	4 5 6	,	7 8 9
millions		thousands		

When saying these greater numbers, I like to think of the commas as having names. The first comma from the right is named "thousand," and the second from the right is "million."

Example 1
Say 123,456,789.

"123 million, 456 thousand, 789" or "one hundred twenty-three million, four hundred fifty-six thousand, seven hundred eighty-nine."

Notice that you never say "and" when reading multiple-digit numbers. This is reserved for the decimal point, which we will study in a succeeding book. Practice saying and writing multiple numbers.

Place-Value Notation

This is a way of writing numbers that emphasizes the place value.

Example 2
Write 8,543,971 with place-value notation.

8,000,000 + 500,000 + 40,000 + 3,000 + 900 + 70 + 1

Weight or Mass – 16 Ounces = 1 Pound

Begin by showing the student three Math-U-See blue 10 bars. The weight of these three bars added together is very close to one ounce. Sixteen ounces is the same as one pound. Four red 100 bars plus five blue 10 bars weigh approximately one pound. This will give the student a feel for one ounce and one pound.

The abbreviation for ounce is ***oz***, and the abbreviation for pound is ***lb***. The abbreviation for pound may puzzle students. It comes from the Latin word for pound, which is "libra."

Example 3
How many ounces are in six pounds?

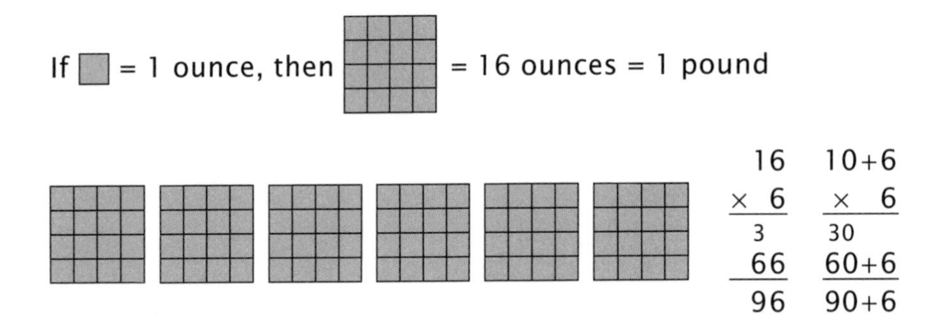

Example 4
How many ounces are in 24 pounds?

16	10+6
×24	× 20+4
2	20
44	40+4
1	100
22	200+20
384	300+80+4

Metric Weight or Mass – 1000 Grams = 1 Kilogram

Metric units of length were mentioned in lesson 14. This is a good time to point out that metric units measure mass, not weight. Two metric units of mass are the gram and the kilogram. One thousand grams make one kilogram. It is helpful to know that a milliliter of water has a mass of about one gram, and a liter of water has a mass of about one kilogram. Also, a small paper clip has a mass of about one gram. The best way for students to become familiar with metric measures at this level is to observe food and beverage labels and make comparisons with other everyday objects.

Mental Math

Here are some more questions to read to your student. These include multiplication and addition. Remember to go slowly at first.

1. One times three, plus nine, equals what number? (12)
2. Four plus six, times five, equals what number? (50)
3. Three times two, plus three, equals what number? (9)
4. Four times eight, plus one, equals what number? (33)
5. Six plus three, times eight, equals what number? (72)
6. Three times three, plus eight, plus one, equals what number? (18)
7. Five plus two, times seven, plus zero, equals what number? (49)
8. Two plus eight, times four, plus five, equals what number? (45)
9. Eight times two, plus one, plus one, equals what number? (18)
10. Two times four, plus one, times five, equals what number? (45)

LESSON 28

More Multiple-Digit Multiplication

There are no new concepts in this lesson, just application of previously-learned concepts to greater numbers. Take your time and be careful with the place value. Estimation will help you recognize errors, especially when using a calculator. If you want to use one to check your answers, this is a good time to do so, but don't let a machine take the place of good thinking!

Example 1

$$\begin{array}{r} 392 \\ \times 147 \\ \hline \end{array} = \begin{array}{r} 392 \\ \times 100 \\ \hline 39{,}200 \end{array} + \begin{array}{r} 392 \\ \times\ 40 \\ \hline {}^{3}\quad\ \ \\ 12{,}680 \\ \hline 15{,}680 \end{array} + \begin{array}{r} 392 \\ \times\ \ 7 \\ \hline {}^{61}\ \ \\ 2{,}134 \\ \hline 2{,}744 \end{array}$$

$$\begin{array}{r} 392 \\ \times 147 \\ \hline {}^{1\ \ 61} \\ 2{,}134 \\ {}^{1\,3}\qquad \\ 12{,}680 \\ 39{,}200 \\ \hline 57{,}624 \end{array} \qquad \begin{array}{r} (4 \\ \times (1 \\ \hline (40{,}000) \end{array} \qquad \begin{array}{r} (400) \\ \times\ (100) \\ \hline (40{,}000) \end{array}$$

You might wonder why the answer is not very close to the estimate. Notice that 147 is rounded to 100, but it is only three away from being 150, which would be rounded up to 200. We can expect the answer to be between 100 times 400 (40,000) and 200 times 400 (80,000). The number 60,000 is between 40,000 and 80,000. Our final answer of 57,624 is fairly close to 60,000.

Example 2

$$\begin{array}{r} 655 \\ \underline{\times 708} \end{array} = \begin{array}{r} 655 \\ \underline{\times 700} \\ {\scriptstyle 33} \\ \underline{425{,}500} \\ 458{,}500 \end{array} + \begin{array}{r} 655 \\ \underline{\times \quad 0} \\ 0 \end{array} + \begin{array}{r} 655 \\ \underline{\times \quad 8} \\ {\scriptstyle 44} \\ \underline{{}^{1}4{,}800} \\ 5{,}240 \end{array}$$

$$\begin{array}{r} 655 \\ \underline{\times 708} \\ {\scriptstyle 44} \\ {}^{1}4{,}800 \\ {\scriptstyle 1\,3\,3} \\ \underline{425{,}500} \\ 463{,}740 \end{array} \qquad \begin{array}{r} (7 \\ \underline{\times\ (7} \\ (490{,}000) \end{array} \qquad \begin{array}{r} (700) \\ \underline{\times\ (700)} \\ (490{,}000) \end{array}$$

Notice that we don't have a separate line for 0 times 655. Since we understand place value, we know that, when multiplying by the hundreds, we begin with the hundreds place. In this case, the number in the hundreds place is the 5 in 425,500. We can also write the two zeros to show that we are in the hundreds place.

Example 3

$$\begin{array}{r} 4{,}382 \\ \underline{\times 961} \end{array} = \begin{array}{r} 4{,}382 \\ \underline{\times 900} \\ {\scriptstyle 1\,2\,7\,1} \\ \underline{3{,}672{,}800} \\ 3{,}943{,}800 \end{array} + \begin{array}{r} 4{,}382 \\ \underline{\times \quad 60} \\ {\scriptstyle 1\,4\,1} \\ \underline{248{,}820} \\ 262{,}920 \end{array} + \begin{array}{r} 4{,}382 \\ \underline{\times \quad 1} \\ 4{,}382 \end{array}$$

$$\begin{array}{r} 4{,}382 \\ \underline{\times 961} \\ {\scriptstyle 2\,1} \\ 4{,}382 \\ {\scriptstyle 2\,{}^{2}1\,4\,1} \\ 248{,}820 \\ {\scriptstyle 1\,2\,7\,1} \\ \underline{3\,672{,}800} \\ 4{,}211{,}102 \end{array} \qquad \begin{array}{r} (4 \\ \underline{\times\ (1,} \\ (4{,}000{,}000) \end{array} \qquad \begin{array}{r} (4{,}000) \\ \underline{\times\ (1{,}000)} \\ (4{,}000{,}000) \end{array}$$

LESSON 29

Prime and Composite Numbers

Multiply by 12

In lesson 26, we learned how to find the factors of a number by building a rectangle and finding its dimensions. In all of the examples, we could build at least two different rectangles with different dimensions, which means there were at least two different sets of factors. Remember that one by six and six by one describe the same rectangle. One may be vertical and the other horizontal, but both of them have the same factors.

A number like 6 or 12, for which there are more than one possible set of factors and from which you can build more than one rectangle, is a ***composite number.*** All the numbers given in lesson 26 were composite numbers. The numbers between 2 and 24 that are composite are 4, 6, 8, 9, 10, 12, 14, 15, 16, 18, 20, 21, 22, and 24.

A whole number greater than one that has only one set of factors or from which you can build only one rectangle is called a ***prime number.*** The definition of a prime number is "any whole number greater than one that has only the factors of one and itself" or "can be divided evenly only by one and itself." The prime numbers between 2 and 24 are 2, 3, 5, 7, 11, 13, 17, 19, and 23. Ask the student to build some of them, and he or she will find that there is only one rectangle to be built and thus only one set of factors. Example 1 illustrates a prime number, and Example 2 illustrates a composite number.

Example 1

Find all possible factors of 13 and tell whether it is prime or composite.

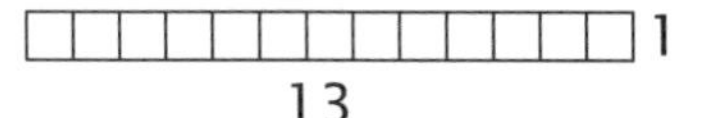

13 has only one set of factors, 1×13, so it is prime.

Example 2
Find all possible factors of 20 and tell whether it is prime or composite.

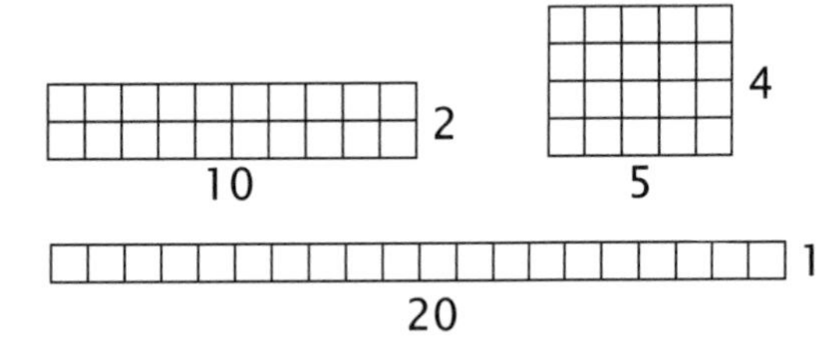

Twenty has three sets of factors, 1×20, 2×10, and 4×5, so it is composite.

Multiplication by 12

Some teachers present multiplying by 12 (the 12 facts) as a set of facts to be memorized. This is a good idea, since we use multiples of 12 in many areas of life. There are 12 items in a dozen, 12 inches in a foot, 12 months in a year, and 12 hours on a clock face. Twelve is also a double-digit number, and with a few strategies we can learn how to multiply by it fairly easily. Practice 1×12 through 12×12 until you can recite the 12 facts quickly.

The number 12 is $10 + 2$, so when you multiply 4×12, think $4 \times (10 + 2)$ and use the Distributive Property: $4 \times 12 = 4(10 + 2) = 4 \times 10 + 4 \times 2 = 40 + 8 = 48$. Take the multiplier times the tens and then times the units and add the answers. You are breaking the problem down into two simpler ones.

Example 3
Solve 8×12.
$8 \times 12 = 8(10 + 2) = 8 \cdot 10 + 8 \cdot 2 = 80 + 16 = 96$
So $8 \times 12 = 96$.

Example 4
Solve 6×12.
$6 \times 12 = 6(10 + 2) = 6 \cdot 10 + 6 \cdot 2 = 60 + 12 = 72$
So $6 \times 12 = 72$.

Example 5
Solve 11×12.

I picture a rectangle over 11 and up 12.
This is 132.

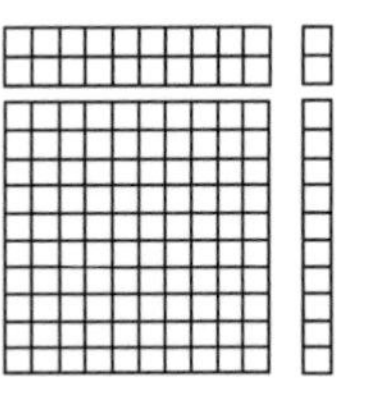

LESSON 30

5,280 Feet = 1 Mile; 2,000 Pounds = 1 Ton

To show how far a mile is, find a straight stretch of road in your area that you can measure with the odometer in your car, or look for the mile markers found on most interstates and turnpikes. If there is an athletic track nearby, find out how long it is and define a mile in terms of how many laps are required. Most tracks are either one fourth or one third of a mile.

There are 5,280 feet in one mile. Since this is true, how many feet are in six miles?

Example 1

How many feet are in six miles? 5,280 × 6 = 31,680 ft

```
  5,280        (5
 ×    6       × (6
 ------       --------
  1 4         (30,000)
 30,280
 ------
 31,680
```

Example 2

How many feet are in 25 miles? 5,280 × 25 = 132,000 ft

```
   5,280        (5,
 ×    25       × (3
 -------       ---------
  1 1 4        (150,000)
  25,000
    1
 104,600
 -------
 132,000
```

Weight - 2,000 Pounds = 1 Ton

One United States ton (sometimes called a "short ton") is equal to 2,000 pounds. One thousand is 10 red hundred blocks. Two thousand is double that, or 20 red hundred blocks.

If □ = 1 pound, then [hundred block] × 20 = 2,000 pounds = 1 ton

Example 3

How many pounds are in four tons? 2,000 × 4 = 8,000 pounds

```
 2,(   )      2,000
 ×    4      ×    4
 8,000        8,000
```

Example 4

How many pounds are in 27 tons? 2,000 × 27 = 54,000 lb

```
  2,000        (2,0
 ×   27       ×   (3
 14,000      (6 0,0 0 0)
 40,000
 54,000
```

Mental Math

Here are some more expressions to read to your student. These include multiplication, addition, and subtraction.

1. Six minus three, plus five, times one, equals what number? (8)
2. Four times three, minus eight, plus seven, equals what number? (11)
3. Seventeen minus nine, plus one, times six, equals what number? (54)
4. Two times two, plus seven, minus one, equals what number? (10)
5. Seven times three, minus one, plus six, equals what number? (26)
6. One times ten, plus eight, minus nine, equals what number? (9)
7. Twelve minus seven, plus four, times three, equals what number? (27)
8. Six times three, plus two, minus ten, equals what number? (10)
9. Five plus zero, times eight, minus twenty, equals what number? (20)
10. Fifteen minus eight, plus three, times ten, equals what number? (100)

APPENDIX A

More on Fractions

The *Gamma* student workbook shows fractions as squares divided into equal parts or shares. The value of using squares to represent fractions will become apparent in *Epsilon*, when operations with fractions are taught in detail.

You may also relate a fraction shown by the square to a fraction number line, which resembles a ruler. Any fraction with a numerator of one is a ***unit fraction.***

Example 1

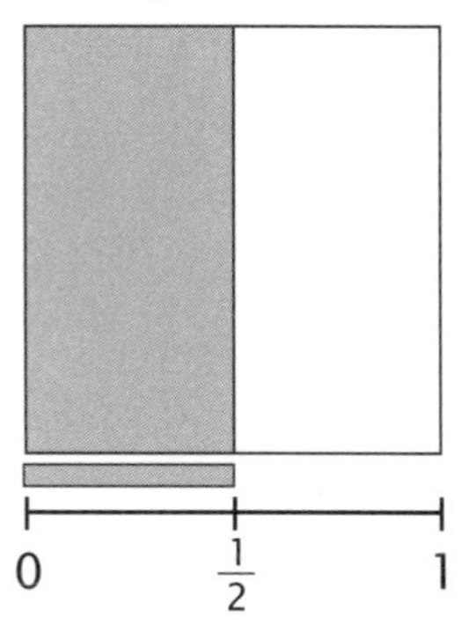

The square and the line are both divided into two parts or shares, so the denominator is 2. One of the parts is shaded, so the numerator is 1. The shaded part is one half of the whole. The shaded line above the ruler represents the **amount** one half, and the vertical line above the 1/2 on the ruler represents the **number** 1/2.

Example 2

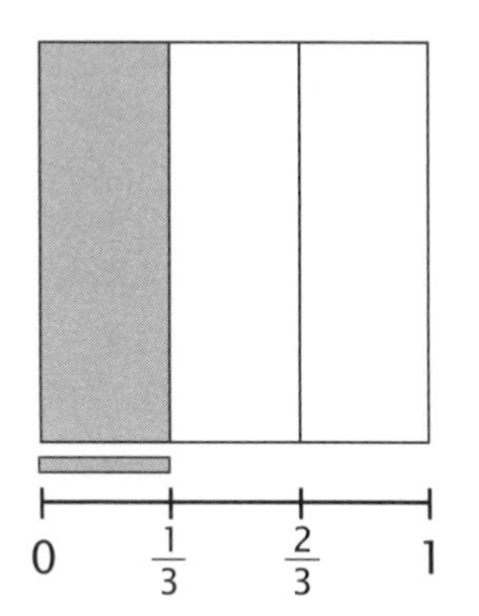

This square shows a share of one third. The shaded line above the ruler represents the amount one third, and the vertical line above the 1/3 on the ruler represents the number 1/3.

Here is another square divided into three parts, but this time two of the parts are shaded, so the fraction that describes the shaded part is 2/3.

Example 3

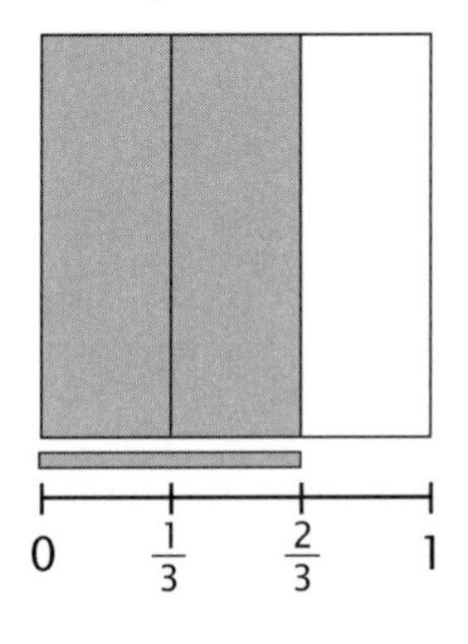

The shaded line above the ruler represents the amount two thirds, and the vertical line above the 2/3 on the ruler represents the number 2/3.

Example 4

This square is one whole. There is only one part, so the denominator is 1. We choose the one part, so we can write the fraction as 1/1. The fraction 1/1 means the same thing as 1.

Example 5

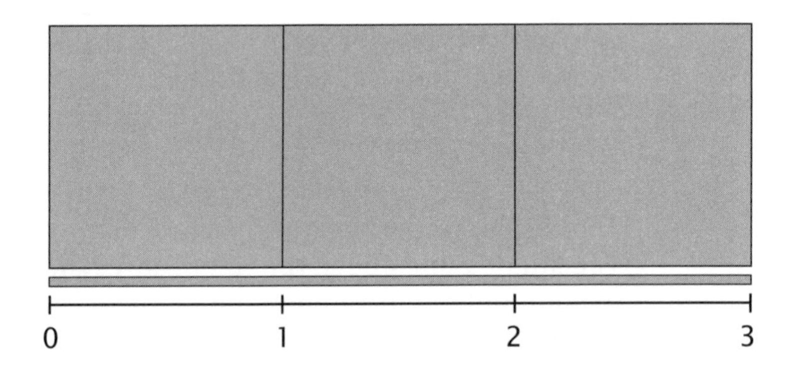

If the whole amount is three, and it is not divided or shared with anyone else: it has only one part. We can write it as 3/1, which means the same as 3.

The fraction squares may also be used to help a student compare two fractions with the same denominator. The diagrams below represent a cake that was cut into five shares. Sam ate 1/5 of the cake, and Tom ate 3/5 of the cake. Who ate the larger piece of cake?

Figure 1

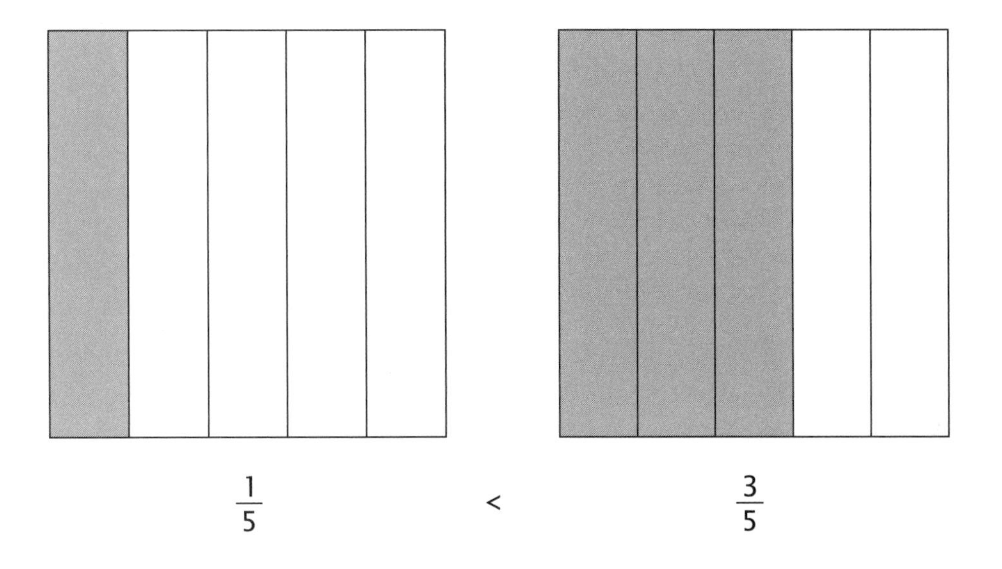

Be sure to stress that the denominators must be the same before you can compare the fractions. Remember, "To compare or combine, you must have the same kind."

As was mentioned, operations with fractions will be taught in detail in the *Epsilon* level of Math-U-See. For now, the best way to give children experience with fractions is to look for real-life examples that involve sharing, measuring for cooking, and using a ruler.

Student Solutions

Lesson Practice 1A

1. done
2. $3 \times 3 = 9$
3. $2 \times 6 = 12$
   $6 \times 2 = 12$
4. $3 \times 4 = 12$
   $4 \times 3 = 12$
5. $5 \times 5 = 25$
6. $2 \times 4 = 8$
   $4 \times 2 = 8$
7. 3×4 rectangle
8. 3×5 rectangle

Lesson Practice 1B

1. $2 \times 2 = 4$
2. $4 \times 6 = 24$
   $6 \times 4 = 24$
3. $2 \times 5 = 10$
   $5 \times 2 = 10$
4. $4 \times 5 = 20$
   $5 \times 4 = 20$
5. $4 \times 7 = 28$
   $7 \times 4 = 28$
6. $2 \times 3 = 6$
   $3 \times 2 = 6$
7. 3×5 rectangle
8. 7×3 rectangle

Lesson Practice 1C

1. $2 \times 7 = 14$
   $7 \times 2 = 14$
2. $3 \times 5 = 15$
   $5 \times 3 = 15$
3. $3 \times 7 = 21$
   $7 \times 3 = 21$
4. $4 \times 4 = 16$
5. $1 \times 8 = 8$
   $8 \times 1 = 8$
6. $4 \times 3 = 12$
   $3 \times 4 = 12$
7. 2×5 rectangle
8. 5×2 rectangle

Systematic Review 1D

1. $5 \times 2 = 10$
   $2 \times 5 = 10$
2. $4 \times 7 = 28$
   $7 \times 4 = 28$
3. $3 + 1 = 4$
4. $4 + 2 = 6$
5. $2 + 8 = 10$
6. $9 + 5 = 14$
7. $8 + 6 = 14$
8. $5 + 5 = 10$
9. $7 + 6 = 13$
10. $4 + 5 = 9$
11. $10 - 1 = 9$
12. $8 - 2 = 6$
13. $15 - 9 = 6$
14. $16 - 8 = 8$
15. $8 - 4 = 4$
16. $7 - 3 = 4$
17. $9 - 6 = 3$
18. $11 - 7 = 4$

Systematic Review 1E

1. $3 \times 2 = 6$
   $2 \times 3 = 6$
2. $5 \times 3 = 15$
   $3 \times 5 = 15$
3. $6 + 1 = 7$
4. $5 + 2 = 7$
5. $2 + 3 = 5$
6. $9 + 7 = 16$
7. $8 + 4 = 12$
8. $6 + 6 = 12$
9. $5 + 6 = 11$
10. $4 + 7 = 11$
11. $8 - 1 = 7$
12. $5 - 2 = 3$

13. $18 - 9 = 9$
14. $14 - 7 = 7$
15. $15 - 8 = 7$
16. $8 - 5 = 3$
17. $9 - 8 = 1$
18. $12 - 4 = 8$

Systematic Review 1F

1. $8 \times 1 = 8$
   $1 \times 8 = 8$
2. $3 \times 3 = 9$
3. $7 + 1 = 8$
4. $6 + 2 = 8$
5. $2 + 9 = 11$
6. $9 + 4 = 13$
7. $8 + 3 = 11$
8. $4 + 4 = 8$
9. $7 + 8 = 15$
10. $5 + 7 = 12$
11. $3 - 1 = 2$
12. $9 - 2 = 7$
13. $16 - 9 = 7$
14. $13 - 8 = 5$
15. $12 - 6 = 6$
16. $7 - 4 = 3$
17. $13 - 7 = 6$
18. $12 - 3 = 9$

Lesson Practice 2A

1. $1 \times 0 = 0$
2. $0 \times 3 = 0$
3. $4 \times 0 = 0$
4. $0 \times 6 = 0$
5. $9 \times 0 = 0$
6. $0 \times 2 = 0$
7. $0 \times 5 = 0$
8. $8 \times 0 = 0$
9. $7 \times 0 = 0$
10. $1 \times 1 = 1$
11. $8 \times 1 = 8$
12. $1 \times 2 = 2$
13. $3 \times 1 = 3$
14. $1 \times 5 = 5$
15. $7 \times 1 = 7$
16. $1 \times 4 = 4$
17. $9 \times 1 = 9$
18. $1 \times 6 = 6$
19. $1 \times 0 = 0$
20. $0 \times 5 = 0$
21. $4 \times 1 = 4$
    $1 \times 4 = 4$
22. $0 + 0 + 0 + 0 + 0 + 0 = 0$
23. 8
24. $3 \times 1 = 3$ flowers
25. $9 \times 1 = 9$ bookmarks

Lesson Practice 2B

1. $5 \times 0 = 0$
2. $0 \times 1 = 0$
3. $6 \times 1 = 6$
4. $1 \times 9 = 9$
5. $1 \times 4 = 4$
6. $7 \times 1 = 7$
7. $1 \times 5 = 5$
8. $3 \times 1 = 3$
9. $1 \times 2 = 2$
10. $8 \times 1 = 8$
11. $1 \times 1 = 1$
12. $7 \times 0 = 0$
13. $8 \times 0 = 0$
14. $0 \times 5 = 0$
15. $0 \times 2 = 0$
16. $0 \times 9 = 0$
17. $0 \times 1 = 0$
18. $3 \times 0 = 0$
19. $0 \times 4 = 0$
20. $6 \times 0 = 0$
21. $5 \times 1 = 5$
    $1 \times 5 = 5$
22. $0 = 0$
23. 10
24. $0 \times 5 = 0$ chores
25. $6 \times 1 = 6$ plates

Lesson Practice 2C

1. $6 \times 0 = 0$
2. $0 \times 2 = 0$
3. $5 \times 1 = 5$
4. $1 \times 8 = 8$
5. $1 \times 7 = 7$
6. $6 \times 1 = 6$
7. $1 \times 2 = 2$
8. $9 \times 1 = 9$
9. $1 \times 5 = 5$
10. $4 \times 1 = 4$
11. $1 \times 0 = 0$
12. $6 \times 0 = 0$
13. $4 \times 0 = 0$
14. $0 \times 3 = 0$
15. $0 \times 8 = 0$
16. $0 \times 2 = 0$
17. $3 \times 1 = 3$
18. $7 \times 0 = 0$
19. $1 \times 4 = 4$
20. $1 \times 8 = 8$
21. $9 \times 1 = 9$
    $1 \times 9 = 9$
22. $0 + 0 + 0 = 0$
23. $9 = 9$
24. $1 \times 8 = 8$ noses
25. $7 \times 0 = 0$ grapes

Systematic Review 2D

1. $1 \times 9 = 9$
2. $3 \times 1 = 3$
3. $7 \times 0 = 0$
4. $9 \times 0 = 0$
5. $1 \times 0 = 0$
6. $1 \times 6 = 6$
7. $1 \times 5 = 5$
8. $1 \times 1 = 1$
9. $0 \times 2 = 0$
10. $0 \times 3 = 0$
11. $0 \times 1 = 0$
12. $7 \times 1 = 7$
13. $4 \times 1 = 4$
14. $2 \times 1 = 2$
15. $0 \times 8 = 0$
16. $6 \times 0 = 0$
17. $5 + 7 = 12$
18. $3 + 4 = 7$
19. $7 - 5 = 2$
20. $5 + 4 = 9$
21. $4 - 0 = 4$
22. $6 + 6 = 12$
23. $6 + 3 = 9$
24. $15 - 9 = 6$
25. $0 + 0 + 0 + 0 + 0 + 0 + 0 = 0$
26. $1 \times 8 = 8$ bunnies

Systematic Review 2E

1. $0 \times 6 = 0$
2. $8 \times 0 = 0$
3. $2 \times 1 = 2$
4. $1 \times 4 = 4$
5. $0 \times 2 = 0$
6. $0 \times 3 = 0$
7. $1 \times 0 = 0$
8. $7 \times 1 = 7$
9. $8 \times 1 = 8$
10. $0 \times 5 = 0$
11. $4 \times 1 = 4$
12. $5 \times 0 = 0$
13. $0 \times 1 = 0$
14. $6 \times 1 = 6$
15. $5 \times 1 = 5$
16. $1 \times 1 = 1$
17. $5 + 3 = 8$
18. $8 + 6 = 14$
19. $15 - 7 = 8$
20. $7 + 3 = 10$
21. $12 - 7 = 5$
22. $7 + 6 = 13$
23. $6 + 5 = 11$
24. $11 - 3 = 8$
25. $3 = 3$
26. $8 \times 1 = 8$ coats

Systematic Review 2F

1. $1 \times 10 = 10$
2. $4 \times 1 = 4$
3. $6 \times 0 = 0$
4. $1 \times 7 = 7$
5. $0 \times 8 = 0$
6. $9 \times 1 = 9$
7. $5 \times 1 = 5$
8. $7 \times 0 = 0$
9. $6 \times 1 = 6$
10. $1 \times 1 = 1$
11. $0 \times 3 = 0$
12. $10 \times 0 = 0$
13. $16 - 8 = 8$
14. $4 + 5 = 9$
15. $7 - 4 = 3$
16. $8 + 2 = 10$
17. $6 + 4 = 10$
18. $12 - 5 = 7$
19. $5 + 6 = 11$
20. $3 + 6 = 9$
21. $13 - 5 = 8$
22. $8 - 0 = 8$
23. $3 + 9 = 12$
24. $17 - 8 = 9$
25. $0 + 0 = 0$
26. $0 \times 4 = 0$ cans

Lesson Practice 3A

1. 2,4,6,8,10,12,
   14,16,18,20
2. 10,20,30,40,50,
   60,70,80,90,100
3. 5,10,15,20,25,
   30,35,40,45,50
4. 5,10,15,20,25,
   30,35,40,45,50
5. 2,4,6,8,10,12,
   14,16,18,20
6. 5,10,15,20,25,
   30,35,40,45,50
7. 10,20,30,40,50,
   60,70,80,90,100

Lesson Practice 3B

1. 2,4,6,8,10,12,
   14,16,18,20
2. 10,20,30,40,50,
   60,70,80,90,100
3. 5,10,15,20,25,
   30,35,40,45,50
4. 5,10,15,20,25,
   30,35,40,45,50
5. 2,4,6,8,10,12,
   14,16,18,20
6. 5,10,15,20,25,
   30,35,40,45,50
7. 10,20,30,40,50,
   60,70,80,90,100

Lesson Practice 3C

1. 2,4,6,8,10,12,
   14,16,18,20
2. 10,20,30,40,50,
   60,70,80,90,100
3. 5,10,15,20,25,
   30,35,40,45,50
4. 5,10,15,20,25,
   30,35,40,45,50
5. 2,4,6,8,10,12,
   14,16,18,20
6. 5,10,15,20,25,
   30,35,40,45,50
7. 10,20,30,40,50,
   60,70,80,90,100

Systematic Review 3D

1. 5, 10, 15, 20, 25, 30, 35, 40, 45, 50
2. 10, 20, 30, 40, 50, 60, 70, 80, 90, 100
3. 2, 4, 6, 8, 10, 12, 14, 16, 18, 20
4. $1 \times 7 = 7$
5. $6 \times 1 = 6$
6. $10 \times 0 = 0$
7. $0 \times 2 = 0$
8. $3 \times 1 = 3$
9. $8 \times 1 = 8$

10. $0 \times 4 = 0$
11. $0 \times 0 = 0$
12. $7 + 8 = 15$
13. $14 - 5 = 9$
14. $12 - 4 = 8$
15. $8 + 9 = 17$
16. $0+0+0+0+0+0+0+0+0 = 0$
17. 5, 10, 15, 20, 25, 30, 35; 7 flowers
18. $7 \times 1 = 7$ cookies
19. $6 + 5 = 11$ chapters
20. $7 + 9 = 16$
    $16 \times 0 = 0$ rocket ships

Systematic Review 3E

1. 5, 10, 15, 20, 25, 30, 35, 40, 45, 50
2. 2, 4, 6, 8, 10, 12, 14, 16, 18, 20
3. 10, 20, 30, 40, 50, 60, 70, 80, 90, 100
4. $5 \times 1 = 5$
5. $6 \times 0 = 0$
6. $1 \times 10 = 10$
7. $9 \times 1 = 9$
8. $0 \times 3 = 0$
9. $1 \times 1 = 1$
10. $0 \times 8 = 0$
11. $4 \times 1 = 4$
12. $8 - 4 = 4$
13. $10 + 3 = 13$
14. $13 - 6 = 7$
15. $5 + 7 = 12$
16. 0
17. 10, 20, 30, 40, 50; 5 bags
18. $7 \times 1 = 7$ times
19. $11 - 6 = 5$ snacks
20. $\$7 + \$3 = \$10$
    $\$10 - \$1 = \$9$

Systematic Review 3F

1. 10, 20, 30, 40, 50, 60, 70, 80, 90, 100
2. 5, 10, 15, 20, 25, 30, 35, 40, 45, 50
3. 2, 4, 6, 8, 10, 12, 14, 16, 18, 20
4. $1 \times 0 = 0$
5. $2 \times 1 = 2$
6. $5 \times 0 = 0$
7. $8 \times 1 = 8$
8. $6 \times 1 = 6$
9. $10 \times 0 = 0$
10. $7 \times 1 = 7$
11. $9 \times 0 = 0$
12. $6 + 7 = 13$
13. $18 - 9 = 9$
14. $13 - 4 = 9$
15. $8 + 5 = 13$
16. $5 = 5$
17. 2, 4, 6, 8, 10, 12, 14, 16, $\underline{18}$ mittens
18. $10 \times 1 = 10$ kites
19. $9 - 2 = 7$
    $7 - 2 = 5$ apples
20. $0 \times 5 = 0$
    $1 \times 6 = 6$
    $0 + 6 = 6$ stories

Lesson Practice 4A

1. $1 \times 2 = 2$
2. $2 \times 3 = 6$
3. $2 \times 5 = 10$
4. $2 \times 7 = 14$
5. $9 \times 2 = 18$
6. $2 \times 2 = 4$
7. $2 \times 10 = 20$
8. $8 \times 2 = 16$
9. $4 \times 2 = 8$
10. $7 \times 2 = 14$
11. $2 \times 6 = 12$
12. $2 \times 0 = 0$
13. $4 \times 2 = 8$
    $2 \times 4 = 8$

14-15. $\frac{0}{2\times0}$ $\frac{2}{2\times1}$ $\frac{4}{2\times2}$ $\frac{6}{2\times3}$ $\frac{8}{2\times4}$ $\frac{10}{2\times5}$
$\frac{12}{2\times6}$ $\frac{14}{2\times7}$ $\frac{16}{2\times8}$ $\frac{18}{2\times9}$ $\frac{20}{2\times10}$

16. $6 \times 2 = 12$ pints
17. 8×2
18. $2+2+2+2+2+2+2+2+2+2 = 20$

19. $10 \times 2 = 20$ pages
20. $7 \times 2 = 14$ pints

Lesson Practice 4B

1. $2 \times 2 = 4$
2. $2 \times 6 = 12$
3. $2 \times 8 = 16$
4. $9 \times 2 = 18$
5. $2 \times 7 = 14$
6. $4 \times 2 = 8$
7. $1 \times 2 = 2$
8. $3 \times 2 = 6$
9. $0 \times 2 = 0$
10. $5 \times 2 = 10$
11. $2 \times 8 = 16$
12. $10 \times 2 = 20$
13. $9 \times 2 = 18$
$2 \times 9 = 18$
14-15. $\frac{0}{2 \cdot 0}$ $\frac{2}{2 \cdot 1}$ $\frac{4}{2 \cdot 2}$ $\frac{6}{2 \cdot 3}$ $\frac{8}{2 \cdot 4}$ $\frac{10}{2 \cdot 5}$
$\frac{12}{2 \cdot 6}$ $\frac{14}{2 \cdot 7}$ $\frac{16}{2 \cdot 8}$ $\frac{18}{2 \cdot 9}$ $\frac{20}{2 \cdot 10}$
16. $8 \times 2 = 16$ pints
17. 6×2
18. $2 + 2 + 2 + 2 + 2 + 2 + 2 = 14$
19. $3 \times 2 = 6$ children
20. $8 \times 2 = 16$ pints

Lesson Practice 4C

1. $9 \times 2 = 18$
2. $2 \times 4 = 8$
3. $2 \times 7 = 14$
4. $8 \times 2 = 16$
5. $5 \times 2 = 10$
6. $1 \times 2 = 2$
7. $2 \times 6 = 12$
8. $10 \times 2 = 20$
9. $2 \times 2 = 4$
10. $3 \times 2 = 6$
11. $2 \times 9 = 18$
12. $0 \times 2 = 0$
13. $2 \times 2 = 4$
14-15. $\frac{0}{(2)(0)}, \frac{2}{(2)(1)}, \frac{4}{(2)(2)}, \frac{6}{(2)(3)},$
$\frac{8}{(2)(4)}, \frac{10}{(2)(5)}, \frac{12}{(2)(6)}, \frac{14}{(2)(7)},$
$\frac{16}{(2)(8)}, \frac{18}{(2)(9)}, \frac{20}{(2)(10)}$
16. $10 \times 2 = 20$ pints
17. 7×2
18. $2 + 2 = 4$
19. $9 \times 2 = 18$ pints
20. $6 \times 2 = 12$ rabbits

Systematic Review 4D

1. $5 \times 2 = 10$
2. $2 \times 6 = 12$
3. $9 \times 2 = 18$
4. $2 \times 8 = 16$
5. $1 \times 3 = 3$
6. $2 \times 7 = 14$
7. $0 \times 6 = 0$
8. $4 \times 2 = 8$
9. $3 \times 2 = 6$
$2 \times 3 = 6$
10. $10 \times 2 = 20$
$2 \times 10 = 20$
11. $2 \times 1 = 2$
$1 \times 2 = 2$
12. $0 \times 2 = 0$
$2 \times 0 = 0$
13. 2, 4, 6, 8, 10,
12, 14, 16, 18, 20
14. $12 - 4 = 8$
15. $9 + 8 = 17$
16. $15 - 7 = 8$
17. $5 + 4 = 9$
18. $500 + 40 + 2$
19. $100 + 60 + 3$
20. $2 + 2 + 2 + 2 + 2 + 2 + 2 = 14$
21. $2 \times 2 = 4$ pints
22. $5 + 3 = 8$ children
$2 \times 8 = 16$ wheels

Systematic Review 4E

1. $0\times2=0$
2. $5\times2=10$
3. $2\times2=4$
4. $2\times4=8$
5. $2\times3=6$
6. $9\times1=9$
7. $2\times6=12$
8. $10\times2=20$
9. $7\times2=14$
   $2\times7=14$
10. $8\times2=16$
    $2\times8=16$
11. $5\times1=5$
    $1\times5=5$
12. $2\times6=12$
    $6\times2=12$
13. 10, 20, 30, 40, 50, 60, 70, 80, 90, 100
14. $16-8=8$
15. $5+3=8$
16. $18-9=9$
17. $7+5=12$
18. $300+50+1$
19. $200+40+9$
20. $7+7=14$
21. $2\times8=16$ mittens
22. $5\times2=10$
    $10-3=7$ eggs

Systematic Review 4F

1. $3\times2=6$
2. $2\times10=20$
3. $8\times2=16$
4. $1\times7=7$
5. $2\times3=6$
6. $2\times6=12$
7. $4\times2=8$
8. $0\times9=0$
9. $5\times2=10$
   $2\times5=10$
10. $7\times2=14$
11. $9\times2=18$
    $2\times9=18$
12. $4\times1=4$
    $1\times4=4$
13. 5, 10, 15, 20, 25, 30, 35, 40, 45, 50
14. $16-9=7$
15. $7+7=14$
16. $9-4=5$
17. $5+6=11$
18. $100+30+1$
19. $400+70+5$
20. $2\times10=20$ peanuts
21. $3\times2=6$ pints
22. $2\times2=4$ hats
    $3\times2=6$ hats
    $4+6=10$ hats

Lesson Practice 5A

1. $10\times0=0$
2. $5\times10=50$
3. $10\times2=20$
4. $6\times10=60$
5. $10\times10=100$
6. $10\times3=30$
7. $10\times9=90$
8. $10\times7=70$
9. $10\times2=20$
10. $10\times5=50$
11. $10\times1=10$
12. $10\times3=30$
13. $10\times7=70$
    $7\times10=70$
14. $4\times10=40$
    $10\times4=40$
15. $10\times6=60$
    $6\times10=60$
16. $10\times3=30$
    $3\times10=30$

17. 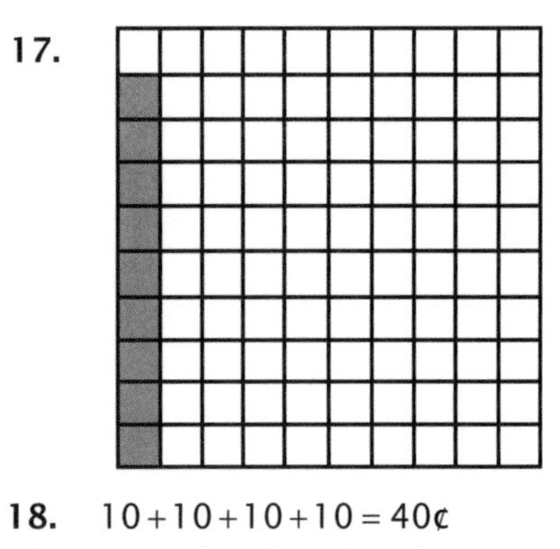

18. $10+10+10+10=40¢$
19. $10+10+10+10+10+$
    $10+10+10+10=90$
20. $10\times6=60$ cars

Lesson Practice 5B

1. $10\times8=80$
2. $1\times10=10$
3. $10\times9=90$
4. $0\times10=0$
5. $10\times5=50$
6. $10\times4=40$
7. $10\times6=60$
8. $10\times10=100$
9. $10\times8=80$
10. $10\times7=70$
11. $10\times2=20$
12. $10\times1=10$
13. $10\times5=50$
    $5\times10=50$
14. $8\times10=80$
    $10\times8=80$
15. $10\times0=0$
    $0\times10=0$
16. $10\times9=90$
    $9\times10=90$
17. $\frac{0}{(10)(0)}$ $\frac{10}{(10)(1)}$ $\frac{20}{(10)(2)}$ $\frac{30}{(10)(3)}$
    $\frac{40}{(10)(4)}$ $\frac{50}{(10)(5)}$ $\frac{60}{(10)(6)}$ $\frac{70}{(10)(7)}$
    $\frac{80}{(10)(8)}$ $\frac{90}{(10)(9)}$ $\frac{100}{(10)(10)}$
18. $10+10+10+10+10+10+10=70¢$
19. $10\times6=60$
20. $10\times5=50$ problems

Lesson Practice 5C

1. $3\times10=30$
2. $8\times10=80$
3. $10\times1=10$
4. $2\times10=20$
5. $10\times9=90$
6. $7\times10=70$
7. $10\times5=50$
8. $6\times10=60$
9. $10\times0=0$
10. $10\times4=40$
11. $10\times10=100$
12. $10\times3=30$
13. $10\times1=10$
    $1\times10=10$
14. $10\times4=40$
    $4\times10=40$
15. $10\times2=20$
    $2\times10=20$
16. $7\times10=70$
    $10\times7=70$
17. see 5A #17
18. $10+10+10+10+10=50¢$
19. $10\times3=30$
20. $\$10\times2=\$20$

Systematic Review 5D

1. $10\times5=50$
2. $7\times10=70$
3. $10\times2=20$
4. $10\times10=100$
5. $2\times5=10$
6. $10\times5=50$
7. $6\times2=12$
8. $7\times2=14$
9. $1\times3=3$
10. $9\times2=18$
11. $10\times8=80$
12. $10\times4=40$

13. $9 \times 2 = 18$
$2 \times 9 = 18$
14. $4 \times 2 = 8$
$2 \times 4 = 8$
15. $10 \times 3 = 30$
$3 \times 10 = 30$
16. $5 \times 2 = 10$
$2 \times 5 = 10$
17. done
18. $\begin{array}{r} 43 \\ +43 \\ \hline 86 \end{array}$
19. $\begin{array}{r} 28 \\ -16 \\ \hline 12 \end{array}$
20. $\begin{array}{r} 89 \\ -51 \\ \hline 38 \end{array}$
21. $7 \times 10 = 70$ hours
22. $70 + 20 = 90$ hours

Systematic Review 5E

1. $10 \times 8 = 80$
2. $6 \times 10 = 60$
3. $10 \times 9 = 90$
4. $10 \times 0 = 0$
5. $5 \times 1 = 5$
6. $6 \times 2 = 12$
7. $8 \times 1 = 8$
8. $10 \times 5 = 50$
9. $2 \times 2 = 4$
10. $2 \times 5 = 10$
11. $9 \times 1 = 9$
$1 \times 9 = 9$
12. $3 \times 10 = 30$
$10 \times 3 = 30$
13. $300 + 80 + 9$
14. $70 + 2$
15. $\begin{array}{r} 46 \\ +22 \\ \hline 68 \end{array}$
16. $\begin{array}{r} 51 \\ +12 \\ \hline 63 \end{array}$
17. $\begin{array}{r} 37 \\ -23 \\ \hline 14 \end{array}$
18. $\begin{array}{r} 94 \\ -43 \\ \hline 51 \end{array}$
19. $10 + 10 + 10 + 10 +$
$10 + 10 + 10 + 10 = 80¢$
20. $4 \times 10 = 40$ fingers
21. $6 + 4 = 10$
$10 \times 10 = 100$ pieces
22. $9 \times 2 = 18$ pints

Systematic Review 5F

1. $4 \times 1 = 4$
2. $2 \times 10 = 20$
3. $10 \times 3 = 30$
4. $10 \times 9 = 90$
5. $6 \times 2 = 12$
6. $2 \times 8 = 16$
7. $10 \times 7 = 70$
8. $10 \times 1 = 10$
9. $3 \times 2 = 6$
10. $4 \times 2 = 8$
11. $1 \times 6 = 6$
12. $9 \times 0 = 0$
13. $100 + 60 + 4$
14. $50 + 8$
15. $\begin{array}{r} 52 \\ -20 \\ \hline 32 \end{array}$
16. $\begin{array}{r} 64 \\ +13 \\ \hline 77 \end{array}$
17. $\begin{array}{r} 35 \\ +34 \\ \hline 69 \end{array}$
18. $\begin{array}{r} 14 \\ -12 \\ \hline 2 \end{array}$

19. $5+5+5+5+5+5+5+5+5+5=50$
20. $9\times10=90¢$
21. Wayne: $\$5\times10=\$50$
Together: $\$50+\$5=\$55$
22. $2\times8=16$ pints

Lesson Practice 6A

1. $5\times4=20$
2. $5\times9=45$
3. $5\times8=40$
4. $5\times10=50$
5. $2\times5=10$
6. $5\times5=25$
7. $5\times1=5$
8. $5\times3=15$
9. $7\times5=35$
10. $0\times5=0$
11. $6\times5=30$
12. $5\times5=25$
13. $5\times10=50$
$10\times5=50$
14. $5\times7=35$
$7\times5=35$
15. $5\times3=15$
$3\times5=15$
16. $5\times6=30$
$6\times5=30$
17.

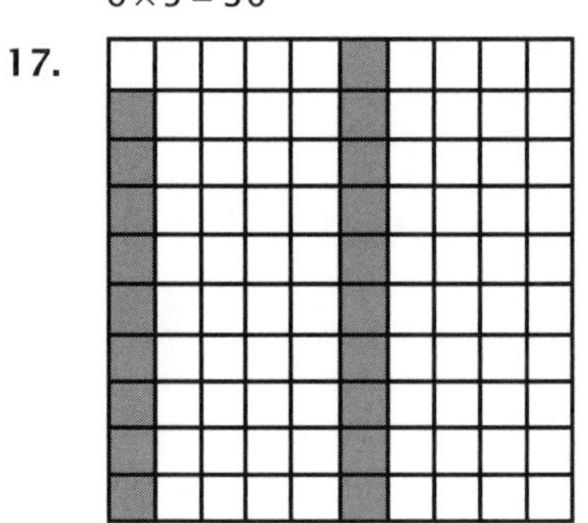

18. $5+5=10¢$
19. $5+5+5+5=20$
20. $5\times8=40$ fingers

Lesson Practice 6B

1. $5\times8=40$
2. $5\times4=20$
3. $5\times6=30$
4. $5\times1=5$
5. $2\times5=10$
6. $9\times5=45$
7. $5\times3=15$
8. $7\times5=35$
9. $5\times5=25$
10. $5\times0=0$
11. $10\times5=50$
12. $5\times4=20$
13. $5\times2=10$
$2\times5=10$
14. $5\times8=40$
$8\times5=40$
15. $9\times5=45$
$5\times9=45$
16. $5\times1=5$
$1\times5=5$
17. $6\times5=30$
$5\times6=30$
18. $\frac{0}{5\times0}$ $\frac{5}{5\times1}$ $\frac{10}{5\times2}$ $\frac{15}{5\times3}$ $\frac{20}{5\times4}$ $\frac{25}{5\times5}$
$\frac{30}{5\times6}$ $\frac{35}{5\times7}$ $\frac{40}{5\times8}$ $\frac{45}{5\times9}$ $\frac{50}{5\times10}$
19. $5+5+5+5+5+5+5+5=40$
20. $5\times5¢=25¢$

Lesson Practice 6C

1. $5\times2=10$
2. $6\times5=30$
3. $5\times10=50$
4. $0\times5=0$
5. $5\times1=5$
6. $7\times5=35$
7. $5\times5=25$
8. $4\times5=20$
9. $5\times3=15$
10. $8\times5=40$
11. $5\times9=45$
12. $5\times6=30$

13. $5 \times 4 = 20$
    $4 \times 5 = 20$
14. $5 \times 10 = 50$
    $10 \times 5 = 50$
15. $5 \times 7 = 35$
    $7 \times 5 = 35$
16. $5 \times 0 = 0$
    $0 \times 5 = 0$
17. see 6A #17
18. $5 \times 8 = 40$¢
19. $5+5+5+5+5+$
    $5+5+5+5+5 = 50$
20. $5 \times 7 = 35$ pages

Systematic Review 6D

1. $5 \times 6 = 30$
2. $10 \times 7 = 70$
3. $2 \times 6 = 12$
4. $5 \times 8 = 40$
5. $9 \times 5 = 45$
6. $10 \times 6 = 60$
7. $8 \times 2 = 16$
8. $7 \times 5 = 35$
9. $0 \times 3 = 0$
10. $9 \times 1 = 9$
11. $5 \times 4 = 20$
12. $10 \times 3 = 30$
13. $3 \times 5 = 15$
    $5 \times 3 = 15$
14. $7 \times 2 = 14$
    $2 \times 7 = 14$
15. $10 \times 5 = 50$
    $5 \times 10 = 50$
16. $1 \times 2 = 2$
    $2 \times 1 = 2$
17. $\begin{array}{r} \scriptstyle 1 \\ 25 \\ +36 \\ \hline 61 \end{array}$
18. $\begin{array}{r} \scriptstyle 1 \\ 178 \\ +34 \\ \hline 112 \end{array}$
19. $\begin{array}{r} \scriptstyle 1 \\ 149 \\ +51 \\ \hline 100 \end{array}$
20. $\begin{array}{r} \scriptstyle 1 \\ 65 \\ +15 \\ \hline 80 \end{array}$
21. 5, 10, 15, 20, 25, 30, 35;
    7 letters
22. $25+58 = 83$ minutes

Systematic Review 6E

1. $5 \times 5 = 25$
2. $1 \times 5 = 5$
3. $2 \times 9 = 18$
4. $10 \times 10 = 100$
5. $10 \times 8 = 80$
6. $5 \times 2 = 10$
7. $6 \times 5 = 30$
8. $9 \times 5 = 45$
9. $7 \times 1 = 7$
10. $2 \times 3 = 6$
11. $8 \times 2 = 16$
12. $9 \times 0 = 0$
13. $4 \times 5 = 20$
    $5 \times 4 = 20$
14. $10 \times 2 = 20$
    $2 \times 10 = 20$
15. $5 \times 7 = 35$
    $7 \times 5 = 35$
16. $5 \times 3 = 15$
    $3 \times 5 = 15$
17. $\begin{array}{r} \scriptstyle 1 \\ 27 \\ +34 \\ \hline 61 \end{array}$
18. $\begin{array}{r} \scriptstyle 1 \\ 19 \\ +13 \\ \hline 32 \end{array}$
19. $\begin{array}{r} \scriptstyle 1 \\ 61 \\ +29 \\ \hline 90 \end{array}$

20. $$\begin{array}{r} ^{1} \\ 47 \\ +37 \\ \hline 84 \end{array}$$

21. $5+5+5=15$
22. $9\times5=45$ miles
23. $18+19=37$ experiments
24. $15-5=10$
    $10+15=25$ daisies

Systematic Review 6F

1. $5\times0=0$
2. $5\times10=50$
3. $8\times5=40$
4. $9\times10=90$
5. $10\times4=40$
6. $2\times6=12$
7. $5\times2=10$
8. $5\times3=15$
9. $6\times0=0$
10. $8\times1=8$
11. $2\times4=8$
12. $5\times7=35$
13. $$\begin{array}{r} 61 \\ -30 \\ \hline 31 \end{array}$$
14. $$\begin{array}{r} ^{1} \\ 28 \\ +23 \\ \hline 51 \end{array}$$
15. $$\begin{array}{r} ^{1} \\ 49 \\ +14 \\ \hline 63 \end{array}$$
16. $$\begin{array}{r} 35 \\ +64 \\ \hline 99 \end{array}$$
17. $$\begin{array}{r} ^{1} \\ 57 \\ +27 \\ \hline 84 \end{array}$$
18. $$\begin{array}{r} 24 \\ -13 \\ \hline 11 \end{array}$$
19. $$\begin{array}{r} 88 \\ -24 \\ \hline 64 \end{array}$$
20. $$\begin{array}{r} ^{1} \\ 83 \\ +\ 9 \\ \hline 92 \end{array}$$
21. $9+9+9+9+9=45$
22. $8\times5=40¢$
23. $8\times2=16$ pints
24. $43+29=72$ lights
    $72-10=62$ bulbs

Lesson Practice 7A

1. done
2. $5\times5=25$
3. $2\times2=4$ sq in
4. $2\times3=6$ sq in
5. $5\times2=10$ sq ft
6. $1\times1=1$ sq mi
7. $5\times8=40$ sq ft
8. $10\times9=90$ tiles

Lesson Practice 7B

1. done
2. $10\times5=50$ sq in
3. $2\times4=8$ sq in
4. $5\times7=35$ sq ft
5. $10\times10=100$ sq ft
6. $2\times9=18$ sq in
7. $5\times5=25$ sq mi
8. $5\times6=30$ sq in

Lesson Practice 7C

1. $6\times1=6$ sq ft
2. $2\times2=4$ sq in
3. $10\times2=20$ sq mi
4. $10\times4=40$ sq ft
5. $5\times9=45$ sq in
6. $5\times4=20$ sq mi
7. 10 ft $\times$ 10 ft $=100$ sq ft; yes
8. $8\times2=16$ blocks

Systematic Review 7D

1. $10 \times 8 = 80$ sq in
2. $3 \times 5 = 15$ sq ft
3. $1 \times 1 = 1$ sq mi
4. $5 \times 9 = 45$
5. $10 \times 6 = 60$
6. $5 \times 7 = 35$
7. $0 \times 6 = 0$
8. $1 \times 8 = 8$
9. $7 \times 2 = 14$
10. $2 \times 5 = 10$
11. $5 \times 6 = 30$
12. $\begin{array}{r} {}^{1}\\ 35 \\ +56 \\ \hline 91 \end{array}$
13. $\begin{array}{r} 48 \\ -24 \\ \hline 24 \end{array}$
14. $\begin{array}{r} {}^{1}\\ 78 \\ +\ 9 \\ \hline 87 \end{array}$
15. $\begin{array}{r} 84 \\ -13 \\ \hline 71 \end{array}$
16. $2 \times 9 = 18$ sq mi
17. $6 \times 2 = 12$ pints
18. $3 \text{ ft} \times 2 \text{ ft} = 6$ sq ft
    $6 \times 2 = 12$ cats

Systematic Review 7E

1. $10 \times 7 = 70$ sq ft
2. $5 \times 5 = 25$ sq mi
3. $2 \times 7 = 14$ sq in
4. $1 \times 9 = 9$
5. $10 \times 4 = 40$
6. $5 \times 8 = 40$
7. $2 \times 6 = 12$
8. $10 \times 1 = 10$
9. $5 \times 3 = 15$
10. $10 \times 7 = 70$
11. $9 \times 2 = 18$
12. $\begin{array}{r} 79 \\ -21 \\ \hline 58 \end{array}$
13. $\begin{array}{r} {}^{1}\\ 32 \\ +59 \\ \hline 91 \end{array}$
14. $\begin{array}{r} 63 \\ -30 \\ \hline 33 \end{array}$
15. $\begin{array}{r} {}^{1}\\ 45 \\ +45 \\ \hline 90 \end{array}$
16. $6 \times 5¢ = 30¢$
17. 3×5 sq ft $= 15$ sq ft
18. $16 + 26 = 42$ acorns
    $42 - 12 = 30$ acorns

Systematic Review 7F

1. $4 \times 1 = 4$ sq in
2. $10 \times 10 = 100$ sq mi
3. $5 \times 9 = 45$ sq units
4. $10 \times 3 = 30$
5. $5 \times 2 = 10$
6. $4 \times 5 = 20$
7. $10 \times 2 = 20$
8. $1 \times 1 = 1$
9. $8 \times 2 = 16$
10. $2 \times 4 = 8$
11. $5 \times 7 = 35$
12. $\begin{array}{r} {}^{1}\\ {}^{1}67 \\ +53 \\ \hline 120 \end{array}$
13. $\begin{array}{r} {}^{1}\\ 19 \\ +12 \\ \hline 31 \end{array}$
14. $\begin{array}{r} {}^{1}\\ 93 \\ +\ 8 \\ \hline 101 \end{array}$

15. $\begin{array}{r} 61 \\ -40 \\ \hline 21 \end{array}$
16. $50+50=100$
    $10\times10=100$ sq ft; yes
17. 5, 10, 15, 20, 25, 30, 35, 40, 45; 9 rows
18. $13+12=25$
    $25-5=20$ pts
    2, 4, 6, 8, 10, 12, 14, 16, 18, 20
    10 qts of jelly left

Lesson Practice 8A

1. done
2. $5\times\underline{5}=25$
3. $10\times\underline{8}=80$
4. $1\times\underline{7}=7$
5. $4\times\underline{5}=20$
6. $10\times\underline{5}=50$
7. $2\times\underline{9}=18$
8. $5\times\underline{3}=15$
9. $3\times\underline{1}=3$
10. $2\times\underline{2}=4$
11. $1\times\underline{10}=10$
12. $8\times\underline{2}=16$
13. $7\times\underline{2}=14$
14. $5\times\underline{6}=30$
15. $10\times\underline{10}=100$
16. $5\times\underline{9}=45$
17. $2\times\underline{6}=12$; 6 people
18. $2\times\underline{9}=18$; 9 gumdrops
19. $10\times\underline{4}=40$; 4 vases
20. $5\times\underline{10}=50$; 10 weeks

Lesson Practice 8B

1. $2\times\underline{10}=20$
2. $5\times\underline{7}=35$
3. $10\times\underline{9}=90$
4. $6\times\underline{0}=0$
5. $4\times\underline{2}=8$
6. $7\times\underline{10}=70$
7. $5\times\underline{6}=30$
8. $8\times\underline{1}=8$
9. $2\times\underline{6}=12$
10. $10\times\underline{6}=60$
11. $8\times\underline{5}=40$
12. $10\times\underline{2}=20$
13. $3\times\underline{2}=6$
14. $1\times\underline{4}=4$
15. $2\times\underline{10}=20$
16. $8\times\underline{0}=0$
17. $2\times\underline{7}=14$; 7 days
18. $2\times\underline{8}=16$; 8 people
19. $5\times\underline{7}=35$; 7 cages
20. $10\times\underline{\$10}=\$100$; \$10 each

Lesson Practice 8C

1. $2\times\underline{6}=12$
2. $5\times\underline{2}=10$
3. $10\times\underline{4}=40$
4. $1\times\underline{6}=6$
5. $5\times\underline{5}=25$
6. $8\times\underline{10}=80$
7. $9\times\underline{1}=9$
8. $5\times\underline{9}=45$
9. $2\times\underline{8}=16$
10. $3\times\underline{5}=15$
11. $10\times\underline{10}=100$
12. $9\times\underline{2}=18$
13. $3\times\underline{10}=30$
14. $5\times\underline{8}=40$
15. $5\times\underline{0}=0$
16. $2\times\underline{1}=2$
17. $2\times\underline{5}=10$; 5 children
18. $5\times\underline{4}=20$; 4 cars
19. $2\times\underline{4}=8$; 4 qt
20. $\$10\times\underline{9}=\$90$; 9 bills

Systematic Review 8D

1. $2\times\underline{10}=20$
2. $10\times\underline{6}=60$
3. $5\times\underline{7}=35$
4. $5\times\underline{1}=5$

5. $7\times\underline{2}=14$
6. $6\times\underline{5}=30$
7. $5\times1=5$ sq in
8. $5\times5=25$ sq ft
9. $5\times4=20$ sq mi
10. $$\begin{array}{r} 1 \\ 61 \\ +29 \\ \hline 90 \end{array}$$
11. $$\begin{array}{r} 1 \\ 72 \\ +38 \\ \hline 110 \end{array}$$
12. $$\begin{array}{r} 44 \\ +55 \\ \hline 99 \end{array}$$
13. $$\begin{array}{r} 86 \\ +73 \\ \hline 159 \end{array}$$
14. $$\begin{array}{r} 2\phantom{\,{}^{1}4} \\ \cancel{3}\,{}^{1}4 \\ -1\;6 \\ \hline 1\;8 \end{array}$$
15. $$\begin{array}{r} 6\phantom{\,{}^{1}4} \\ \cancel{7}\,{}^{1}4 \\ -3\;8 \\ \hline 3\;6 \end{array}$$
16. $$\begin{array}{r} 6\phantom{\,{}^{1}1} \\ \cancel{7}\,{}^{1}1 \\ -59 \\ \hline 12 \end{array}$$
17. $$\begin{array}{r} 67 \\ -25 \\ \hline 42 \end{array}$$
18. $65-37=28$ cookies

Systematic Review 8E

1. $5\times\underline{8}=40$
2. $10\times\underline{7}=70$
3. $2\times\underline{9}=18$
4. $1\times\underline{1}=1$
5. $3\times\underline{2}=6$
6. $9\times\underline{5}=45$
7. $10\times4=40$ sq in
8. $2\times2=4$ sq ft
9. $10\times8=80$ sq ft
10. $$\begin{array}{r} 1 \\ 35 \\ +65 \\ \hline 100 \end{array}$$
11. $$\begin{array}{r} 53 \\ +61 \\ \hline 114 \end{array}$$
12. $$\begin{array}{r} 1 \\ 99 \\ +22 \\ \hline 121 \end{array}$$
13. $$\begin{array}{r} 1 \\ 26 \\ +48 \\ \hline 74 \end{array}$$
14. $$\begin{array}{r} 17 \\ -\;\;9 \\ \hline 8 \end{array}$$
15. $$\begin{array}{r} 3\phantom{\,{}^{1}5} \\ \cancel{4}\,{}^{1}5 \\ -2\;6 \\ \hline 1\;9 \end{array}$$
16. $$\begin{array}{r} {}^{8}\cancel{9}\,{}^{1}3 \\ -3\;4 \\ \hline 5\;9 \end{array}$$
17. $$\begin{array}{r} 52 \\ -42 \\ \hline 10 \end{array}$$
18. $5\times2=10$ pints
19. $2\times\underline{6}=12$; 6 peanuts
20. $10\times10=100$ sq mi

Systematic Review 8F

1. $4\times\underline{5}=20$
2. $2\times\underline{2}=4$
3. $10\times\underline{2}=20$
4. $5\times\underline{5}=25$
5. $9\times\underline{0}=0$
6. $3\times\underline{10}=30$
7. $4\times2=8$ sq mi
8. $1\times1=1$ sq in
9. $5\times6=30$ sq ft

10. $\begin{array}{r} 21 \\ +58 \\ \hline 79 \end{array}$

11. $\begin{array}{r} {\scriptstyle 1} \\ 15 \\ +17 \\ \hline 32 \end{array}$

12. $\begin{array}{r} {\scriptstyle 1} \\ 84 \\ +46 \\ \hline 130 \end{array}$

13. $\begin{array}{r} {\scriptstyle 1} \\ 76 \\ +75 \\ \hline 151 \end{array}$

14. $\begin{array}{r} {\scriptstyle 1}\phantom{^{1}4} \\ \cancel{2}\,{}^{1}4 \\ -\ 1\ 5 \\ \hline 9 \end{array}$

15. $\begin{array}{r} {\scriptstyle 4}\phantom{^{1}3} \\ \cancel{5}\,{}^{1}3 \\ -\ 3\ 5 \\ \hline 1\ 8 \end{array}$

16. $\begin{array}{r} {\scriptstyle 5}\phantom{^{1}5} \\ \cancel{6}\,{}^{1}5 \\ -\ 1\ 9 \\ \hline 4\ 6 \end{array}$

17. $\begin{array}{r} {\scriptstyle 2}\phantom{^{1}7} \\ \cancel{3}\,{}^{1}7 \\ -\ \ \ 8 \\ \hline 2\ 9 \end{array}$

18. $68+94=162$ blocks
19. $3\times\underline{\$5}=\$15$; $5 each
20. $6\times2=12$ pints

Lesson Practice 9A

1. 9, 18, 27, 36, 45, 54, 63, 72, 81, 90
2. 9, 18, 27, 36, 45, 54, 63, 72, 81, 90
3. $\frac{2}{9},\frac{4}{18},\frac{6}{27},\frac{8}{36},\frac{10}{45},\frac{12}{54},\frac{14}{63},\frac{16}{72},\frac{18}{81},\frac{20}{90}$
4. 9, 18, 27, 36, 45, $\underline{54}$ players
5. 9, 18, 27, 36, 45, 54, 63, 72, 81, $\underline{90}$¢
6. 9, 18, $\underline{27}$ slices

Lesson Practice 9B

1. 9, 18, 27, 36, 45, 54, 63, 72, 81, 90
2. 9, 18, 27, 36, 45, 54, 63, 72, 81, 90
3. $\frac{5}{9},\frac{10}{18},\frac{15}{27},\frac{20}{36},\frac{25}{45},\frac{30}{54},\frac{35}{63},\frac{40}{72},\frac{45}{81},\frac{50}{90}$
4. 9, 18, 27, $\underline{36}$ children
5. 9, 18, 27, 36, 45, 54, 63, $\underline{72}$ rocks
6. 9, 18, 27, 36, 45, 54, 63, 72, 81, $\underline{90}$ sq ft

Lesson Practice 9C

1. 9, 18, 27, 36, 45, 54, 63, 72, 81, 90
2. 9, 18, 27, 36, 45, 54, 63, 72, 81, 90
3. $\frac{9}{10},\frac{18}{20},\frac{27}{30},\frac{36}{40},\frac{45}{50},\frac{54}{60},$ $\frac{63}{70},\frac{72}{80},\frac{81}{90},\frac{90}{100}$
4. 9, 18, 27, 36, 45, 56, 63, 72, $\underline{81}$ tiles
5. 9, 18, 27, 36, $\underline{45}$ cities
6. 9, 18, 27, 36, 45, 54, $\underline{63}$ letters

Systematic Review 9D

1. 9, 18, 27, 36, 45, 54, 63, 72, 81, 90
2. 5, 10, 15, 20, 25, 30, 35, 40, 45, 50
3. $5\times\underline{5}=25$
4. $6\times\underline{10}=60$
5. $2\times\underline{9}=18$
6. $1\times7=7$
7. $5\times4=20$
8. $10\times8=80$
9. $7\times2=14$ sq in
10. $10\times10=100$ sq mi
11. $5\times3=15$ sq in
12. $\begin{array}{r} 25 \\ +62 \\ \hline 87 \end{array}$
13. $\begin{array}{r} {\scriptstyle 8}\phantom{^{1}1} \\ \cancel{9}\,{}^{1}1 \\ -\ 4\ 5 \\ \hline 4\ 6 \end{array}$
14. $\begin{array}{r} {\scriptstyle 1} \\ 16 \\ +17 \\ \hline 33 \end{array}$

15. $\begin{array}{r} 86 \\ -73 \\ \hline 13 \end{array}$

16. $3 \times 5 = 15¢$

17. $4 \times 5 = 20¢$

18. $10 \times 8 = 80¢$

19. $14 + 29 = 43$ arrows

20. $9, \underline{18}$ players

Systematic Review 9E

1. 9, 18, 27, 36, 45, 54, 63, 72, 81, 90

2. $\frac{5}{10}, \frac{10}{20}, \frac{15}{30}, \frac{20}{40}, \frac{25}{50}, \frac{30}{60},$
$\frac{35}{70}, \frac{40}{80}, \frac{45}{90}, \frac{50}{100}$

3. $3 \times \underline{10} = 30$

4. $4 \times \underline{2} = 8$

5. $5 \times \underline{7} = 35$

6. $0 \times 8 = 0$

7. $9 \times 10 = 90$

8. $6 \times 5 = 30$

9. $10 \times 1 = 10$ sq ft

10. $5 \times 5 = 25$ sq in

11. $5 \times 2 = 10$ sq in

12. $\begin{array}{r} {}^{1} \\ 13 \\ +47 \\ \hline 60 \end{array}$

13. $\begin{array}{r} \overset{6}{\cancel{7}}\,{}^{1}5 \\ -1\;8 \\ \hline 5\;7 \end{array}$

14. $\begin{array}{r} {}^{1} \\ 43 \\ +59 \\ \hline 102 \end{array}$

15. $\begin{array}{r} \overset{8}{\cancel{9}}\,{}^{1}4 \\ -5\;5 \\ \hline 3\;9 \end{array}$

16. 9, 18, 27, 36, 45, 54; 6 vans

17. $7 \times 10 = 70¢$

18. $5 \times \underline{9} = 45$; 9 rings

19. $44 + 26 = 70$
$70 - 12 = 58$ tons

20. $6 \times 2 = 12$ jars

Systematic Review 9F

1. 9, 18, 27, 36, 45, 54, 63, 72, 81, 90

2. $\frac{2}{(2)(1)}, \frac{4}{(2)(2)}, \frac{6}{(2)(3)}, \frac{8}{(2)(4)}, \frac{10}{(2)(5)},$
$\frac{12}{(2)(6)}, \frac{14}{(2)(7)} \frac{16}{(2)(8)}, \frac{18}{(2)(9)}, \frac{20}{(2)(10)}$

3. $8 \times \underline{5} = 40$

4. $9 \times \underline{10} = 90$

5. $1 \times \underline{4} = 4$

6. $9 \times 1 = 9$

7. $2 \times 10 = 20$

8. $2 \times 9 = 18$

9. $3 \times 2 = 6$ sq mi

10. $2 \times 2 = 4$ sq ft

11. $9 \times 5 = 45$ sq in

12. $\begin{array}{r} \overset{1}{\cancel{2}}\,{}^{1}7 \\ -1\;9 \\ \hline 8 \end{array}$

13. $\begin{array}{r} {}^{1} \\ 61 \\ +39 \\ \hline 100 \end{array}$

14. $\begin{array}{r} {}^{1} \\ 54 \\ +47 \\ \hline 101 \end{array}$

15. $\begin{array}{r} \overset{7}{\cancel{8}}\,{}^{1}2 \\ -7\;3 \\ \hline 9 \end{array}$

16. 9, 18, 27, 36, 45, 54, 63, 72, 81; 9 days

17. $6 \times 5 = 30¢$

18. $6 \times 10 = 60¢$
$60¢ + 30¢ = 90¢$

19. $11 + 9 = 20$ pails for Jack
$20 - 6 = 14$ pails

20. $7 \times 10 = 70$ books

Lesson Practice 10A

1. $9 \times 4 = 36$
2. $9 \times 9 = 81$
3. $9 \times 8 = 72$
4. $10 \times 9 = 90$
5. $2 \times 9 = 18$
6. $5 \times 9 = 45$
7. $9 \times 1 = 9$
8. $9 \times 3 = 27$
9. $9 \times 7 = 63$
10. $9 \times 9 = 81$
11. $6 \times 9 = 54$
12. $3 \times 9 = 27$
13. $9 \times 8 = 72$
    $8 \times 9 = 72$
14. $9 \times 4 = 36$
    $4 \times 9 = 36$
15. $9 \times 5 = 45$
    $5 \times 9 = 45$
16. $9 \times 7 = 63$
    $7 \times 9 = 63$
17. 0
18. $9+9+9+9+9+9+9+9+9+9 = 90$
19. $9+9 = 18$
20. $9+9+9+9+9+9 = 54$
21. 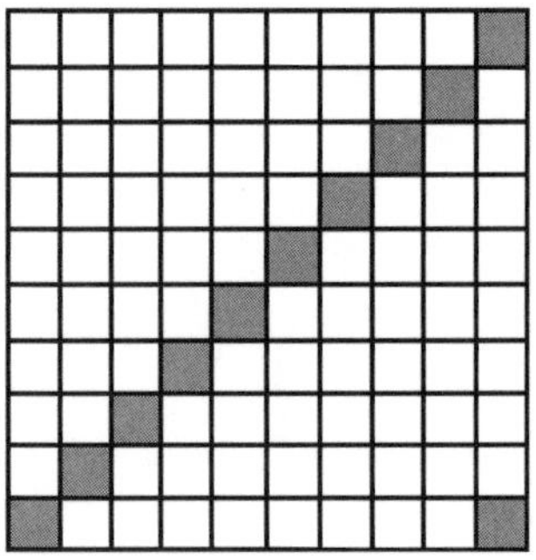
22. $9 \times 4 = 36$

Lesson Practice 10B

1. $9 \times 10 = 90$
2. $9 \times 6 = 54$
3. $9 \times 2 = 18$
4. $0 \times 9 = 0$
5. $7 \times 9 = 63$
6. $3 \times 9 = 27$
7. $9 \times 5 = 45$
8. $9 \times 8 = 72$
9. $4 \times 9 = 36$
10. $1 \times 9 = 9$
11. $9 \times 9 = 81$
12. $6 \times 9 = 54$
13. $9 \times 3 = 27$
    $3 \times 9 = 27$
14. $9 \times 2 = 18$
    $2 \times 9 = 18$
15. $9 \times 6 = 54$
    $6 \times 9 = 54$
16. $9 \times 10 = 90$
    $10 \times 9 = 90$
17. $9 \times 7 = 63$
18. $9 \times 5 = 45$
19. $9 \times 8 = 72$
20. $9 \times 4 = 36$
21. $\frac{0}{9 \cdot 0}, \frac{9}{9 \cdot 1}, \frac{18}{9 \cdot 2}, \frac{27}{9 \cdot 3}, \frac{36}{9 \cdot 4}, \frac{45}{9 \cdot 5}, \frac{54}{9 \cdot 6}, \frac{63}{9 \cdot 7}, \frac{72}{9 \cdot 8}, \frac{81}{9 \cdot 9}, \frac{90}{9 \cdot 10}$
22. $9 \times 5 = 45$ planets
23. $9 \times 9 = 81$ squares
24. $9 \times 3 = 27$ acres

Lesson Practice 10C

1. $9 \times 1 = 9$
2. $9 \times 8 = 72$
3. $9 \times 6 = 54$
4. $10 \times 2 = 20$
5. $9 \times 9 = 81$
6. $3 \times 9 = 27$
7. $5 \times 9 = 45$
8. $9 \times 7 = 63$
9. $9 \times 0 = 0$
10. $4 \times 9 = 36$
11. $10 \times 9 = 90$
12. $9 \times 8 = 72$
13. $9 \times 1 = 9$
    $1 \times 9 = 9$
14. $9 \times 7 = 63$
    $7 \times 9 = 63$

15. $9 \times 4 = 36$
$4 \times 9 = 36$
16. $9 \times 5 = 45$
$5 \times 9 = 45$
17. $9 \times 3 = 27$
18. $9 \times 9 = 81$
19. $9 \times 10 = 90$
20. $9 \times 0 = 0$
21.

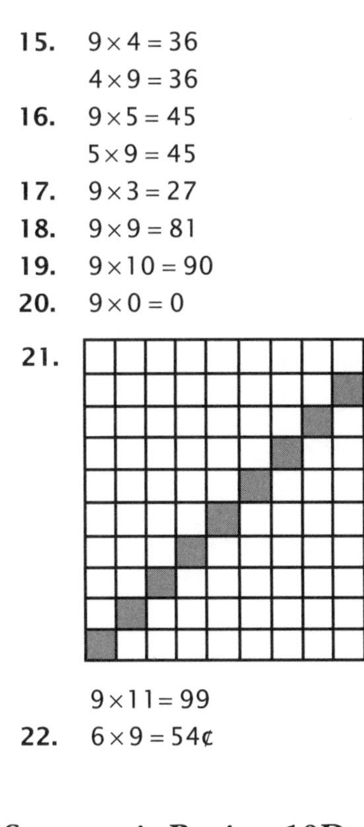

$9 \times 11 = 99$
22. $6 \times 9 = 54¢$

Systematic Review 10D

1. $9 \times 5 = 45$
2. $7 \times 9 = 63$
3. $5 \times 6 = 30$
4. $8 \times 2 = 16$
5. $10 \times 9 = 90$
6. $8 \times 5 = 40$
7. $3 \times 2 = 6$
8. $7 \times 5 = 35$
9. $9 \times \underline{6} = 54$
10. $2 \times \underline{7} = 14$
11. $10 \times \underline{5} = 50$
12. $6 \times \underline{0} = 0$
13. $9 \times 9 = 81$ sq in
14. $3 \times 9 = 27$ sq ft
15. $4 \times 5 = 20$ sq mi
16. $8 \times \$9 = \$72$; \$9 per gift
17. $9 \times 4 = 36$
$36 + 15 = 51$ min
18. $5 \times 5 = 25$
$3 \times 10 = 30$
$30 + 25 = 55¢$
19. $35 - 28 = 7$ questions
20. $2 \times 6 = 12$
$53 - 12 = 41$ sq in

Systematic Review 10E

1. $2 \times 9 = 18$
2. $5 \times 5 = 25$
3. $7 \times 10 = 70$
4. $6 \times 9 = 54$
5. $1 \times 9 = 9$
6. $9 \times 9 = 81$
7. $5 \times 3 = 15$
8. $9 \times 5 = 45$
9. $2 \times \underline{10} = 20$
10. $7 \times \underline{9} = 63$
11. $2 \times \underline{5} = 10$
12. $3 \times \underline{9} = 27$
13. $9 \times 4 = 36$ sq in
14. $2 \times 7 = 14$ sq ft
15. $1 \times 1 = 1$ sq in
16. $\frac{0}{9 \times 0}, \frac{9}{9 \times 1}, \frac{18}{9 \times 2}, \frac{27}{9 \times 3}, \frac{36}{9 \times 4}, \frac{45}{9 \times 5}, \frac{54}{9 \times 6}, \frac{63}{9 \times 7}, \frac{72}{9 \times 8}, \frac{81}{9 \times 9}, \frac{90}{9 \times 10}$
17. $6 \times 10 = 60¢$
$7 \times 5 = 35¢$
$60 + 35 = 95¢$
18. $9 \times 8 = 72$ sq mi
19. $13 + 12 = 25$
$25 - 7 = 18$ pints
20. $45 + 39 = 84$ miles

Systematic Review 10F

1. $2 \times 4 = 8$
2. $5 \times 6 = 30$
3. $9 \times 5 = 45$
4. $3 \times 9 = 27$
5. $8 \times 9 = 72$
6. $9 \times 6 = 54$
7. $10 \times 3 = 30$
8. $4 \times 5 = 20$
9. $4 \times \underline{9} = 36$
10. $2 \times \underline{9} = 18$

11. $8 \times \underline{10} = 80$
12. $8 \times \underline{5} = 40$
13. $9 \times 7 = 63$ sq in
14. $10 \times 10 = 100$ sq ft
15. $2 \times 6 = 12$ sq ft
16. $7 \times 5 = 35¢$ - Jim
    $4 \times 10 = 40¢$ - Lisa
    Lisa has more.
    $35 + 40 = 75¢$
17. $9 \times 9 = 81$
    $81 - 3 = 78$ clubs
18. $17 + 25 = 42$ children
19. $53 - 45 = 8$ minutes
20. $5 \times 9 = 45$ tiles

Lesson Practice 11A

1. 3, 6, 9, 12, 15, 18, 21, 24, 27, 30
2. 3, 6, 9, 12, 15, 18, 21, 24, 27, 30
3. 3, 6, 9, 12, 15, 18, 21, 24, 27, 30
4. $\frac{2}{3} = \frac{4}{6} = \frac{6}{9} = \frac{8}{12} = \frac{10}{15} = \frac{12}{18} = \frac{14}{21} = \frac{16}{24} = \frac{18}{27} = \frac{20}{30}$
5. 3, 6, 9, 12, $\underline{15}$ wheels
6. 3, 6, 9, 12, 15, $\underline{18}$ years
7. 3, 6, 9, 12, 15, 18, 21, $\underline{24}$ sides
8. 3, 6, 9, $\underline{12}$ dots

Lesson Practice 11B

1. 3, 6, 9, 12, 15, 18, 21, 24, 27, 30
2. 3, 6, 9, 12, 15, 18, 21, 24, 27, 30
3. 3, 6, 9, 12, 15, 18, 21, 24, 27, 30
4. $\frac{3}{5} = \frac{6}{10} = \frac{9}{15} = \frac{12}{20} = \frac{15}{25} = \frac{18}{30} = \frac{21}{35} = \frac{24}{40} = \frac{27}{45} = \frac{30}{50}$
5. 3, 6, $\underline{9}$ sides
6. 3, 6, 9, 12, 15, 18, 21, 24, $\underline{27}$ baby steps
7. 3, 6, 9, 12, 15, 18, 21, $\underline{24}$ people
8. 3, $\underline{6}$ hands

Lesson Practice 11C

1. 3, 6, 9, 12, 15, 18, 21, 24, 27, 30
2. 3, 6, 9, 12, 15, 18, 21, 24, 27, 30
3. 3, 6, 9, 12, 15, 18, 21, 24, 27, 30
4. $\frac{3}{10} = \frac{6}{20} = \frac{9}{30} = \frac{12}{40} = \frac{15}{50} = \frac{18}{60} = \frac{21}{70} = \frac{24}{80} = \frac{27}{90} = \frac{30}{100}$
5. 3, 6, 9, $\underline{12}$ shirts
6. 3, 6, 9, 12, 15, 18, $\underline{21}$ books
7. 3, 6, 9, 12, 15, 18, 21, 24, 27, $\underline{\$30}$
8. 3, $\underline{6}$ things

Systematic Review 11D

1. 3, 6, 9, 12, 15, 18, 21, 24, 27, 30
2. $\frac{0}{(9)(0)}, \frac{9}{(9)(1)}, \frac{18}{(9)(2)}, \frac{27}{(9)(3)}, \frac{36}{(9)(4)}, \frac{45}{(9)(5)}, \frac{54}{(9)(6)}, \frac{63}{(9)(7)}, \frac{72}{(9)(8)}, \frac{81}{(9)(9)}, \frac{90}{(9)(10)}$
3. $9 \times \underline{3} = 27$
4. $5 \times \underline{4} = 20$
5. $2 \times \underline{0} = 0$
6. $10 \times \underline{5} = 50$
7. $2 \times 8 = 16$
8. $1 \times 7 = 7$
9. $7 \times 10 = 70$
10. $9 \times 4 = 36$
11. $9 \times 6 = 54$ sq in
12. $5 \times 5 = 25$ sq mi
13. $7 \times 5 = 35$ sq in
14. $$\begin{array}{r} {}^{1} \\ 23 \\ 26 \\ +37 \\ \hline 86 \end{array}$$
15. $$\begin{array}{r} {}^{1} \\ 12 \\ 59 \\ +31 \\ \hline 102 \end{array}$$

16. $\begin{array}{r} {\scriptstyle 1} \\ 15 \\ 15 \\ 44 \\ +24 \\ \hline 98 \end{array}$

17. $\begin{array}{r} {\scriptstyle 2} \\ 34 \\ 56 \\ 11 \\ +\ \ 9 \\ \hline 110 \end{array}$

18. $3+5+4+5+6+2=25$ pies

Systematic Review 11E

1. 3, 6, 9, 12, 15, 18, 21, 24, 27, 30
2. $\frac{0}{5\cdot 0}, \frac{5}{5\cdot 1}, \frac{10}{5\cdot 2}, \frac{15}{5\cdot 3}, \frac{20}{5\cdot 4}, \frac{25}{5\cdot 5}, \frac{30}{5\cdot 6}, \frac{35}{5\cdot 7}, \frac{40}{5\cdot 8}, \frac{45}{5\cdot 9}, \frac{50}{5\cdot 10}$
3. $7\times\underline{9}=63$
4. $6\times\underline{5}=30$
5. $2\times\underline{4}=8$
6. $6\times\underline{10}=60$
7. $3\times 5=15$
8. $8\times 9=72$
9. $9\times 5=45$
10. $8\times 1=8$
11. $8\times 5=40$ sq in
12. $2\times 2=4$ sq mi
13. $9\times 3=27$ sq in
14. $\begin{array}{r} {\scriptstyle 1} \\ 45 \\ 31 \\ +15 \\ \hline 91 \end{array}$
15. $\begin{array}{r} {\scriptstyle 1} \\ 26 \\ 84 \\ +32 \\ \hline 142 \end{array}$
16. $\begin{array}{r} {\scriptstyle 1} \\ 23 \\ 15 \\ 17 \\ +\ \ 2 \\ \hline 57 \end{array}$
17. $\begin{array}{r} {\scriptstyle 2} \\ 31 \\ 29 \\ 32 \\ +\ \ 8 \\ \hline 100 \end{array}$
18. $9\times 10=90¢$
$2\times 5=10¢$
$90+10=100¢=\$1.00$
19. 3, 6, 9, 12, 15, $\underline{18}$ sides
20. $52-35=\underline{17}$ minutes

Systematic Review 11F

1. 3, 6, 9, 12, 15, 18, 21, 24, 27, 30
2. $\frac{0}{2\times 0}, \frac{2}{2\times 1}, \frac{4}{2\times 2}, \frac{6}{2\times 3}, \frac{8}{2\times 4}, \frac{10}{2\times 5}, \frac{12}{2\times 6}, \frac{14}{2\times 7}, \frac{16}{2\times 8}, \frac{18}{2\times 9}, \frac{20}{2\times 10}$
3. $2\times\underline{3}=6$
4. $4\times\underline{10}=40$
5. $6\times\underline{9}=54$
6. $5\times\underline{5}=25$
7. $9\times 9=81$
8. $4\times 9=36$
9. $5\times 7=35$
10. $6\times 0=0$
11. $9\times 7=63$ sq ft
12. $1\times 1=1$ sq ft
13. $6\times 2=12$ sq in
14. $\begin{array}{r} {\scriptstyle 1} \\ 19 \\ 91 \\ +\ \ 7 \\ \hline 117 \end{array}$
15. $\begin{array}{r} {\scriptstyle 1} \\ 17 \\ 36 \\ +44 \\ \hline 97 \end{array}$

16.

$$\begin{array}{r} 1 \\ 55 \\ 41 \\ 65 \\ +\ 2 \\ \hline 163 \end{array}$$

17.

$$\begin{array}{r} 1 \\ 51 \\ 12 \\ 24 \\ +38 \\ \hline 125 \end{array}$$

18. $5 \times \underline{9} = 45¢$; 9 nickels

19. 3, 6, 9, 12, 15, 18, $\underline{21}$ dots

20. $\$50 - \$38 = \$12$

$3 \times \underline{4} = \$12$; $4 each

Lesson Practice 12A

1. $3 \times 4 = 12$
2. $9 \times 3 = 27$
3. $3 \times 7 = 21$
4. $10 \times 3 = 30$
5. $3 \times 7 = 21$
6. $1 \times 3 = 3$
7. $3 \times 4 = 12$
8. $8 \times 3 = 24$
9. $8 \times 3 = 24$
10. $3 \times 5 = 15$
11. $3 \times 3 = 9$
12. $6 \times 3 = 18$
13. 3, $\underline{6}$
14. 3, 6, 9, 12, 15, $\underline{18}$
15. (See lesson 12 in instruction manual.)
16. $\frac{0}{(3)(0)}, \frac{3}{(3)(1)}, \frac{6}{(3)(2)}, \frac{9}{(3)(3)}, \frac{12}{(3)(4)}, \frac{15}{(3)(5)}, \frac{18}{(3)(6)}, \frac{21}{(3)(7)}, \frac{24}{(3)(8)}, \frac{27}{(3)(9)}, \frac{30}{(3)(10)}$
17. $\frac{3}{5} = \frac{6}{10} = \frac{9}{15} = \frac{12}{20} = \frac{15}{25} = \frac{18}{30} = \frac{21}{35} = \frac{24}{40} = \frac{27}{45} = \frac{30}{50}$
18. $4 \times 3 = 12$ ft
19. $6 \times 3 = 18$ tsp
20. $3 \times 3 = 9$ ft

Lesson Practice 12B

1. $3 \times 10 = 30$
2. $3 \times 3 = 9$
3. $3 \times 5 = 15$
4. $6 \times 3 = 18$
5. $3 \times 2 = 6$
6. $0 \times 3 = 0$
7. $3 \times 9 = 27$
8. $7 \times 3 = 21$
9. $1 \times 3 = 3$
10. $3 \times 8 = 24$
11. $4 \times 3 = 12$
12. $10 \times 3 = 30$
13. 3, 6, 9, 12, 15, 18, 21, 24, $\underline{27}$
14. 3, 6, 9, 12, $\underline{15}$
15. $3 \times 6 = 18$

    $6 \times 3 = 18$
16. $3 \times 2 = 6$

    $2 \times 3 = 6$
17. $3 \times 8 = 24'$
18. $7 \times 3 = 21$
19. $4 \times 3 = 12$
20. $5 \times 3 = 15$ pillows

Lesson Practice 12C

1. $3 \times 1 = 3$
2. $3 \times 5 = 15$
3. $3 \times 10 = 30$
4. $2 \times 3 = 6$
5. $3 \times 3 = 9$
6. $9 \times 3 = 27$
7. $3 \times 6 = 18$
8. $4 \times 3 = 12$
9. $7 \times 3 = 21$
10. $3 \times 0 = 0$
11. $8 \times 3 = 24$
12. $5 \times 3 = 15$
13. 3, 6, 9, 12, 15, 18, 21, $\underline{24}$
14. 3, 6, $\underline{9}$

15. 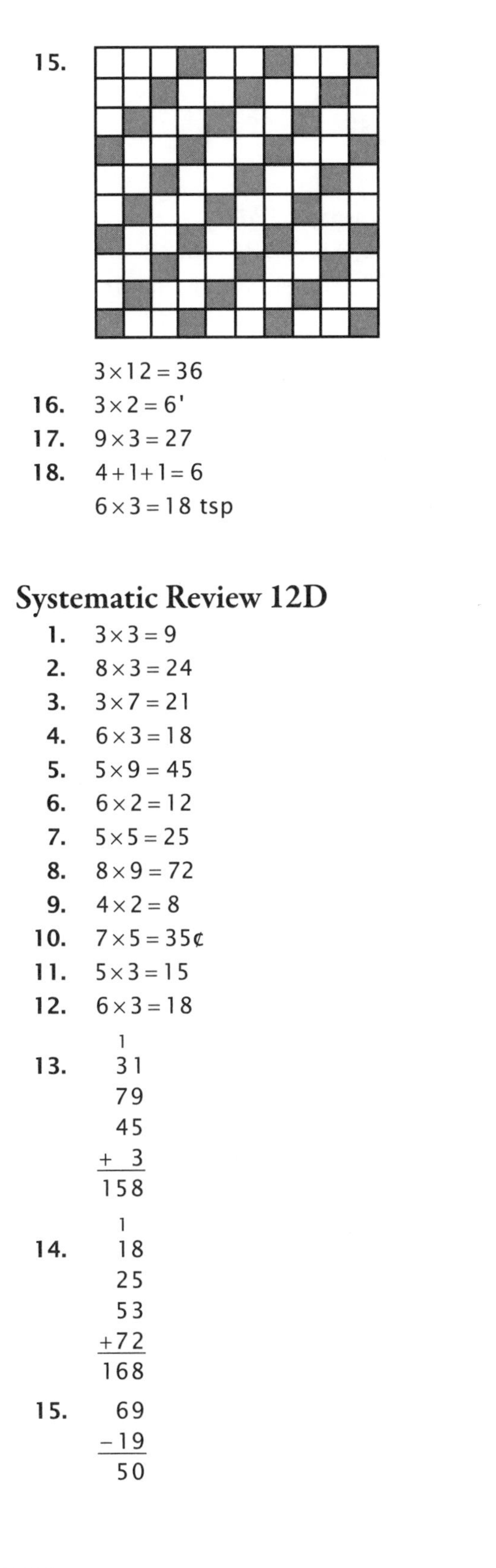

$3 \times 12 = 36$

16. $3 \times 2 = 6'$
17. $9 \times 3 = 27$
18. $4 + 1 + 1 = 6$

$6 \times 3 = 18$ tsp

Systematic Review 12D

1. $3 \times 3 = 9$
2. $8 \times 3 = 24$
3. $3 \times 7 = 21$
4. $6 \times 3 = 18$
5. $5 \times 9 = 45$
6. $6 \times 2 = 12$
7. $5 \times 5 = 25$
8. $8 \times 9 = 72$
9. $4 \times 2 = 8$
10. $7 \times 5 = 35¢$
11. $5 \times 3 = 15$
12. $6 \times 3 = 18$
13. $$\begin{array}{r} ^{1} \\ 31 \\ 79 \\ 45 \\ +\ \ 3 \\ \hline 158 \end{array}$$
14. $$\begin{array}{r} ^{1} \\ 18 \\ 25 \\ 53 \\ +72 \\ \hline 168 \end{array}$$
15. $$\begin{array}{r} 69 \\ -19 \\ \hline 50 \end{array}$$

16. $$\begin{array}{r} {}^{7}\not{8}\,{}^{1}1 \\ -27 \\ \hline 54 \end{array}$$
17. $3 \times \underline{4} = 12$; 4 yards
18. $3 \times 9 = 27$ children
19. $5 + 13 + 7 + 15 = 40$ leaves
20. $\$8 + \$18 = \$26$

$\$26 - \$7 = \$19$

Systematic Review 12E

1. $3 \times 5 = 15$
2. $10 \times 3 = 30$
3. $3 \times 6 = 18$
4. $8 \times 3 = 24$
5. $5 \times 8 = 40$
6. $7 \times 9 = 63$
7. $10 \times 6 = 60$
8. $8 \times 2 = 16$
9. $7 \times 2 = 14$
10. $4 \times 10 = 40$
11. $9 \times 3 = 27$
12. $3 \times 3 = 9$
13. $$\begin{array}{r} ^{1} \\ 57 \\ 21 \\ 22 \\ +15 \\ \hline 115 \end{array}$$
14. $$\begin{array}{r} ^{1} \\ 43 \\ 44 \\ 63 \\ +11 \\ \hline 161 \end{array}$$
15. $$\begin{array}{r} {}^{4}\not{5}\,{}^{1}3 \\ -24 \\ \hline 29 \end{array}$$
16. $$\begin{array}{r} {}^{5}\not{6}\,{}^{1}5 \\ -19 \\ \hline 46 \end{array}$$
17. $2 \times 3 = 6$ tsp

$\underline{2 \text{ Tbsp}} \times 3 = 6$ tsp

18. $3 \times 3 = 9$ apples
$3 \times 9 = 27$ worms
19. $12 + 25 + 3 + 10 = 50$ miles
20. $3 \times 7 = 21$ sq yd
$21 - 11 = 10$ sq yd

Systematic Review 12F

1. $4 \times 3 = 12$
2. $7 \times 3 = 21$
3. $3 \times 8 = 24$
4. $1 \times 3 = 3$
5. $10 \times 3 = 30$
6. $6 \times 3 = 18$
7. $9 \times 6 = 54$
8. $5 \times 4 = 20$
9. $9 \times 2 = 18$
10. $6 \times 5 = 30$
11. $3 \times 3 = 9$
12. $5 \times 3 = 15$
13. $$\begin{array}{r} {}^{1} \\ 17 \\ 16 \\ 12 \\ +14 \\ \hline 59 \end{array}$$
14. $$\begin{array}{r} {}^{1} \\ 35 \\ 74 \\ 15 \\ +24 \\ \hline 148 \end{array}$$
15. $$\begin{array}{r} \overset{3}{\cancel{4}}{}^{1}1 \\ -32 \\ \hline 9 \end{array}$$
16. $$\begin{array}{r} \overset{2}{\cancel{3}}{}^{1}8 \\ -29 \\ \hline 9 \end{array}$$
17. $3 \times \underline{10} = 30$; 10 yards
18. $3 \times 9 = 27$
$30 - 27 = \$3$
19. $57 + 14 = 71$
$71 - 71 = 0$ dogs
20. $4 \times 10 = 40¢$
$3 \times 5 = 15¢$
$15 + 40 + 11 = 66¢$

Lesson Practice 13A

1. 6, 12, 18, 24, 30, 36, 42, 48, 54, 60
2. 6, 12, 18, 24, 30, 36, 42, 48, 54, 60
3. 6, 12, 18, 24, 30, 36, 42, 48, 54, 60
4. $\frac{1}{2} = \frac{2}{4} = \frac{3}{6} = \frac{4}{8} = \frac{5}{10}$
5. $\frac{1}{3} = \frac{2}{6} = \frac{3}{9} = \frac{4}{12} = \frac{5}{15}$
6. 6, 12, $\underline{18}$ paintings
7. 6, 12, 18, 24, 30, 36, 42, $\underline{48}$ crackers

Lesson Practice 13B

1. 6, 12, 18, 24, 30, 36, 42, 48, 54, 60
2. 6, 12, 18, 24, 30, 36, 42, 48, 54, 60
3. 6, 12, 18, 24, 30, 36, 42, 48, 54, 60
4. $\frac{1}{6} = \frac{2}{12} = \frac{3}{18} = \frac{4}{24} = \frac{5}{30}$
5. $\frac{2}{6} = \frac{4}{12} = \frac{6}{18} = \frac{8}{24} = \frac{10}{30}$
6. 6, $\underline{12}$ cones
7. 6, 12, 18, 24, 30, $\underline{36}$ people

Lesson Practice 13C

1. 6, 12, 18, 24, 30, 36, 42, 48, 54, 60
2. 6, 12, 18, 24, 30, 36, 42, 48, 54, 60
3. 6, 12, 18, 24, 30, 36, 42, 48, 54, 60
4. $\frac{5}{6} = \frac{10}{12} = \frac{15}{18} = \frac{20}{24} = \frac{25}{30}$
5. $\frac{2}{3} = \frac{4}{6} = \frac{6}{9} = \frac{8}{12} = \frac{10}{15}$
6. 6, 12, 18, 24, 30, 36, 42, 48, $\underline{54}$ times
7. 6, 12, 18, 24, $\underline{30}$ fingers

Systematic Review 13D

1. 6, 12, 18, 24, 30, 36, 42, 48, 54, 60
2. $\frac{3}{5} = \frac{6}{10} = \frac{9}{15} = \frac{12}{20} = \frac{15}{25}$
3. $9 \times 3 = 27$
4. $2 \times 6 = 12$

5. $3 \times 4 = 12$
6. $5 \times 5 = 25$
7. $8 \times 9 = 72$
8. $6 \times 5 = 30$
9. $3 \times 7 = 21$
10. $10 \times 8 = 80$
11. $\begin{array}{r} 73 \\ +45 \\ \hline 118 \end{array}$
12. $\begin{array}{r} \overset{1}{3}8 \\ +67 \\ \hline 105 \end{array}$
13. $\begin{array}{r} \overset{4}{\cancel{5}}\,{}^{1}4 \\ -25 \\ \hline 29 \end{array}$
14. $\begin{array}{r} \overset{7}{\cancel{8}}\,{}^{1}8 \\ -19 \\ \hline 69 \end{array}$
15. done
16. $3+3+3+3=12$ mi
17. $10+12+10+12=44'$
18. $25+25+25+25=100'$
19. 6, 12, 18, 24, 30, <u>36</u> eggs
20. 6, 12, <u>18</u> ft

Systematic Review 13E

1. 6, 12, 18, 24, 30, 36, 42, 48, 54, 60
2. $\frac{2}{5} = \frac{4}{10} = \frac{6}{15} = \frac{8}{20} = \frac{10}{25}$
3. $2 \times 7 = 14$
4. $3 \times 0 = 0$
5. $9 \times 9 = 81$
6. $5 \times 3 = 15$
7. $3 \times 3 = 9$
8. $9 \times 4 = 36$
9. $10 \times 7 = 70$
10. $8 \times 2 = 16$
11. $\begin{array}{r} 61 \\ +22 \\ \hline 83 \end{array}$
12. $\begin{array}{r} \overset{1}{4}5 \\ +\ 9 \\ \hline 54 \end{array}$
13. $\begin{array}{r} \overset{6}{\cancel{7}}\,{}^{1}6 \\ -38 \\ \hline 38 \end{array}$
14. $\begin{array}{r} \overset{8}{\cancel{9}}\,{}^{1}3 \\ -44 \\ \hline 49 \end{array}$
15. $16+20+16+20=72$ yd
16. $7+4+7+4=22'$
17. $5+5+5+5=20''$
18. $6+4+6+4=20$ mi
19. 6,12, 18, 24, 30, 36, 40, <u>48</u> ft
20. $8 \times 3 = 24$ tsp

Systematic Review 13F

1. 9, 18, 27, 36, 45, 54, 63, 72, 81, 90
2. $\frac{3}{6} = \frac{6}{12} = \frac{9}{18} = \frac{12}{24} = \frac{15}{30}$
3. $3 \times 9 = 27$
4. $9 \times 2 = 18$
5. $2 \times 4 = 8$
6. $10 \times 6 = 60$
7. $9 \times 5 = 45$
8. $5 \times 4 = 20$
9. $9 \times 6 = 54$
10. $7 \times 9 = 63$
11. $\begin{array}{r} \overset{1}{1}7 \\ +18 \\ \hline 35 \end{array}$
12. $\begin{array}{r} \overset{1}{3}4 \\ +79 \\ \hline 113 \end{array}$
13. $\begin{array}{r} 48 \\ -27 \\ \hline 21 \end{array}$

14. $\begin{array}{r} \overset{4}{\cancel{5}}\,{}^{1}3 \\ -\,1\,9 \\ \hline 3\,4 \end{array}$
15. $11+14+11+14=50$ mi
16. $12+21+12+21=66'$
17. $10+10+10+10=40"$
18. $5+1+5+1=12"$
19. $5\times1=5$ sq in
20. $5\times2=10$
    $10\times9=90$ fingers

Lesson Practice 14A

1. $6\times9=54$
2. $7\times6=42$
3. $10\times6=60$
4. $4\times6=24$
5. $5\times6=30$
6. $1\times6=6$
7. $6\times3=18$
8. $6\times2=12$
9. $9\times6=54$
10. $6\times4=24$
11. $6\times8=48$
12. $7\times6=42$
13. $6\times4=24$
14. $6\times1=6$
15. $6\times5=30$
16. $6\times3=18$
17. 6, 12, 18, 24, 30, 36, 42, <u>48</u>
18. 6, 12, 18, 24, 30, <u>36</u>
19. $6\times11=66$
    $6\times12=72$
20. $6\times7=42$ legs

Lesson Practice 14B

1. $6\times0=0$
2. $6\times6=36$
3. $2\times6=12$
4. $8\times6=48$
5. $3\times6=18$
6. $5\times6=30$
7. $6\times9=54$
8. $6\times10=60$
9. $6\times6=36$
10. $6\times1=6$
11. $6\times4=24$
12. $2\times6=12$
13. $6\times7=42$
14. $6\times5=30$
15. $6\times3=18$
16. $9\times6=54$
17. (See lesson 14 in instruction manual.)
18. $\frac{0}{6\cdot0},\frac{6}{6\cdot1},\frac{12}{6\cdot2},\frac{18}{6\cdot3},\frac{24}{6\cdot4},\frac{30}{6\cdot5},$
    $\frac{36}{6\cdot6},\frac{42}{6\cdot7},\frac{48}{6\cdot8},\frac{54}{6\cdot9},\frac{60}{6\cdot10}$
19. $6\times10=60$
20. $6\times4=24$ wheels

Lesson Practice 14C

1. $6\times10=60$
2. $6\times8=48$
3. $6\times6=36$
4. $4\times6=24$
5. $7\times6=42$
6. $9\times6=54$
7. $6\times5=30$
8. $6\times3=18$
9. $6\times0=0$
10. $6\times6=36$
11. $6\times2=12$
12. $8\times6=48$
13. $6\times5=30$
14. $4\times6=24$
15. $6\times1=6$
16. $7\times6=42$

17. $\frac{0}{6\times0}, \frac{6}{6\times1}, \frac{12}{6\times2}, \frac{18}{6\times3}, \frac{24}{6\times4}, \frac{30}{6\times5}, \frac{36}{6\times6}, \frac{42}{6\times7}, \frac{48}{6\times8}, \frac{54}{6\times9}, \frac{60}{6\times10}$
18. $6\times9=54$
19. $6\times6=36$ feet
20. $6\times6=36$ apartments

Systematic Review 14D

1. $10\times7=70$
2. $9\times6=54$
3. $5\times0=0$
4. $6\times3=18$
5. $3\times9=27$
6. $5\times4=20$
7. $4\times6=24$
8. $7\times9=63$
9. $6\times5=30$
10. $6\times6=36$
11. $9\times5=45$
12. $7\times2=14$
13. $14+2+14+2=32"$
14. $8+10+8=26"$
15. $7+7+7+7=28'$
16. $\frac{1}{6}=\frac{2}{12}=\frac{3}{18}=\frac{4}{24}=\frac{5}{30}$
17. $6\times7=42$ glasses
18. $5\times10=50$
    $3\times5=15$
    $50+15+8=73¢$
19. $50-23=27$ apples
20. $4+8=12$
    $12-6=6$ people

Systematic Review 14E

1. $3\times9=27$
2. $6\times9=54$
3. $5\times5=25$
4. $9\times8=72$
5. $4\times3=12$
6. $6\times8=48$
7. $9\times0=0$
8. $7\times5=35$
9. $3\times6=18$
10. $9\times9=81$
11. $1\times8=8$
12. $6\times7=42$
13. $7+9+7+9=32"$
14. $5+6+3=14'$
15. $9+9+9+9=36$ mi
16. $\frac{3}{5}=\frac{6}{10}=\frac{9}{15}=\frac{12}{20}=\frac{15}{25}$
17. $8\times6=48$ nails
18. $4+16+7+19+6=52$ cards
19. $10\times3=30'$
20. $21+30+21+30=102'$

Systematic Review 14F

1. $4\times9=36$
2. $3\times3=9$
3. $6\times6=36$
4. $7\times6=42$
5. $9\times7=63$
6. $3\times8=24$
7. $9\times5=45$
8. $4\times6=24$
9. $8\times2=16$
10. $9\times3=27$
11. $6\times3=18$
12. $6\times5=30$
13. $17+4+17+4=42"$
14. $7+8+8=23'$
15. $4+4+4+4=16$ yd
16. $\frac{2}{3}=\frac{4}{6}=\frac{6}{9}=\frac{8}{12}=\frac{10}{15}$
17. $5\times3=15$ tsp
18. $\underline{4}\times5¢=20¢$; 4 nickels
19. $3\times4=12$
    $7\times2=14$
    $12+14=26$ feet
20. $3\times6=18$ sq ft

Lesson Practice 15A

1. 4, 8, 12, 16, 20, 24, 28, 32, 36, 40
2. 4, 8, 12, 16, 20, 24, 28, 32, 36, 40
3. 4, 8, 12, 16, 20, 24, 28, 32, 36, 40
4. $\frac{4}{5}=\frac{8}{10}=\frac{12}{15}=\frac{16}{20}=\frac{20}{25}=$
   $\frac{24}{30}=\frac{28}{35}=\frac{32}{40}=\frac{36}{45}=\frac{40}{50}$
5. 4, 8, 12, 16, 20, 24, 28, $\underline{32}$ qt
6. 4, 8, $\underline{12}$
   $12+12=24$ socks
7. 4, 8, 12, 16, 20, 24, 28, 32, $\underline{36}$ people
8. 4, 8, 12, 16, $\underline{20}$ jugs

Lesson Practice 15B

1. 4, 8, 12, 16, 20, 24, 28, 32, 36, 40
2. 4, 8, 12, 16, 20, 24, 28, 32, 36, 40
3. 4, 8, 12, 16, 20, 24, 28, 32, 36, 40
4. $\frac{4}{9}=\frac{8}{18}=\frac{12}{27}=\frac{16}{36}=\frac{20}{45}=$
   $\frac{24}{54}=\frac{28}{63}=\frac{32}{72}=\frac{36}{81}=\frac{40}{90}$
5. 4, 8, 12, 16, 20, 24, 28, 32, 36, $\underline{40}$ qt
6. 4, 8, 12, $\underline{16}$ tires
7. 4, 8, 12, 16, 20, $\underline{24}$ laps
8. 4, $\underline{8}$ qt

Lesson Practice 15C

1. 4, 8, 12, 16, 20, 24, 28, 32, 36, 40
2. 4, 8, 12, 16, 20, 24, 28, 32, 36, 40
3. 4, 8, 12, 16, 20, 24, 28, 32, 36, 40
4. $\frac{2}{4}=\frac{4}{8}=\frac{6}{12}=\frac{8}{16}=\frac{10}{20}=$
   $\frac{12}{24}=\frac{14}{28}=\frac{16}{32}=\frac{18}{36}=\frac{20}{40}$
5. 4, 8, 12, 16, 20, 24, $\underline{28}$ qt
6. 4, 8, $\underline{12}$ pictures
7. 4, 8, 12, 16, 20, 24, 28, $\underline{32}$ flowers
8. 4, 8, 12, 16, 20, 24, 28, 32, $\underline{36}$ jars

Systematic Review 15D

1. 4, 8, 12, 16, 20, 24, 28, 32, 36, 40
2. $\frac{3}{4},\frac{6}{8},\frac{9}{12},\frac{12}{16},\frac{15}{20}$
3. $10\times7=70$
4. $9\times6=54$
5. $5\times0=0$
6. $6\times3=18$
7. $9\times3=27$
8. $5\times4=20$
9. $6\times4=24$
10. $7\times9=63$
11. $5\times\underline{6}=30$
12. $6\times\underline{3}=18$
13. $9\times\underline{5}=45$
14. $6\times\underline{6}=36$
15. done
16. done
17. $10\times10=100$ sq ft
18. $10+10+10+10=40$ ft
19. 4, $\underline{8}$ rings
20. 4, 8, 12, $\underline{16}$ qt

Systematic Review 15E

1. 3, 6, 9, 12, 15, 18, 21, 24, 27, 30
2. $\frac{1}{4}=\frac{2}{8}=\frac{3}{12}=\frac{4}{16}=\frac{5}{20}$
3. $2\times7=14$
4. $9\times3=27$
5. $9\times6=54$
6. $5\times5=25$
7. $9\times8=72$
8. $4\times3=12$
9. $6\times8=48$
10. $9\times0=0$
11. $9\times\underline{0}=0$
12. $7\times\underline{5}=35$
13. $8\times\underline{9}=72$
14. $9\times\underline{3}=27$
15. $3\times7=21$ sq ft
16. $3+7+3+7=20$ ft
17. $9\times9=81$ sq mi
18. $9+9+9+9=36$ mi

19. 6, 12, 18, $\underline{24}$ feet
20. $3\times4=12$ ft
    $3\times6=18$ ft
    $3\times9=27$ ft
    $12+18+27=57$ ft

Systematic Review 15F

1. 4, 8, 12, 16, 20, 24, 28, 32, 36, 40
2. $\frac{2}{5}=\frac{4}{10}=\frac{6}{15}=\frac{8}{20}=\frac{10}{25}$
3. $8\times1=8$
4. $7\times6=42$
5. $4\times9=36$
6. $3\times3=9$
7. $6\times6=36$
8. $7\times6=42$
9. $7\times9=63$
10. $3\times8=24$
11. $3\times\underline{6}=18$
12. $6\times\underline{4}=24$
13. $2\times\underline{8}=16$
14. $9\times\underline{9}=81$
15. $5\times6=30$ sq ft
16. $5+6+5+6=22$ ft
17. $2\times10=20$ sq yd
18. $2+10+2+10=24$ yd
19. 4, 8, 12, 16, 20, 24, $\underline{28}$ days
20. $2\times3=6$
    $2\times7=14$
    $2\times2=4$
    $2\times5=10$
    $6+14+4+10=34$ pints

Lesson Practice 16A

1. $4\times9=36$
2. $7\times4=28$
3. $10\times4=40$
4. $4\times6=24$
5. $5\times4=20$
6. $1\times4=4$
7. $4\times3=12$
8. $4\times2=8$
9. $4\times4=16$
10. $4\times8=32$
11. $4\times5=20$
12. $6\times4=24$
13. $4\times9=36$
14. $4\times4=16$
15. 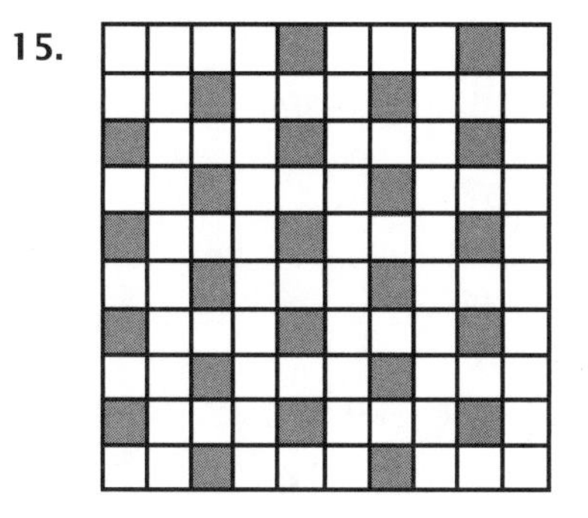

    $4\times11=44$
    $4\times12=48$
16. $4\times6=24$ quarters
17. $7\times4=28$ corners
18. $3\times4=12$ quarters

Lesson Practice 16B

1. $4\times0=0$
2. $4\times10=40$
3. $4\times3=12$
4. $6\times4=24$
5. $2\times4=8$
6. $4\times4=16$
7. $4\times7=28$
8. $9\times4=36$
9. $1\times4=4$
10. $4\times5=20$
11. $4\times8=32$
12. $3\times4=12$
13. 4, 8, 12, 16, 20, 24, 28, 32, 36, 40
14. $\frac{0}{(4)(0)},\frac{4}{(4)(1)},\frac{8}{(4)(2)},\frac{12}{(4)(3)},\frac{16}{(4)(4)},\frac{20}{(4)(5)},$
    $\frac{24}{(4)(6)},\frac{28}{(4)(7)},\frac{32}{(4)(8)},\frac{36}{(4)(9)},\frac{40}{(4)(10)}$
15. $4\times4=16$ quarters
16. $4\times7=28$
17. $4\times9=36$ plates
18. $4\times5=20$ horseshoes

Lesson Practice 16C

1. $4\times 2=8$
2. $4\times 4=16$
3. $6\times 4=24$
4. $4\times 10=40$
5. $1\times 4=4$
6. $4\times 5=20$
7. $4\times 7=28$
8. $3\times 4=12$
9. $8\times 4=32$
10. $4\times 9=36$
11. $4\times 6=24$
12. $4\times 4=16$
13. 4, 8, 12, 16, 20, 24, 28, 32, 36, 40
14. $\frac{0}{4\cdot 0},\frac{4}{4\cdot 1},\frac{8}{4\cdot 2},\frac{12}{4\cdot 3},\frac{16}{4\cdot 4},\frac{20}{4\cdot 5},$
    $\frac{24}{4\cdot 6},\frac{28}{4\cdot 7},\frac{32}{4\cdot 8},\frac{36}{4\cdot 9},\frac{40}{4\cdot 10}$
15. $4\times 8=32$ quarters
16. $4\times 6=24$
17. $D\times 4=12$
    $D=3$
18. $10\times 4=40$ quarters
19. $4\times 9=36$ quarters
20. $5+2=7$
    $7\times 4=28$ qt
    You could also solve this by multiplying first to change gallons to quarts.

Systematic Review 16D

1. $4\times 4=16$
2. $2\times 6=12$
3. $6\times 3=18$
4. $10\times 3=30$
5. $5\times 9=45$
6. $4\times 6=24$
7. $2\times 7=14$
8. $3\times 3=9$
9. $4\times 9=36$
10. $5\times 3=15$
11. $7\times 6=42$
12. $9\times 3=27$
13. done
14. $3\times \underline{3}=9$
15. $4\times 4=16$ sq in
16. $4+4+4+4=16$ in
17. $4\times 8=32$ sq mi
18. $8+4+8+4=24$ mi

Systematic Review 16E

1. $6\times 9=54$
2. $7\times 4=28$
3. $8\times 6=48$
4. $10\times 5=50$
5. $4\times 3=12$
6. $9\times 7=63$
7. $1\times 0=0$
8. $9\times 4=36$
9. $1\times 1=1$
10. $6\times 7=42$
11. $9\times 9=81$
12. $4\times 8=32$
13. $5\times \underline{5}=25$
14. $\underline{4}\times 10=40$
15. $6\times 6=36$ sq in
16. $6+6+6+6=24$ in
17. $4\times 7=28$ sq mi
18. $4+7+4+7=22$ mi
19. $6+2=8$ horses
    $8\times 4=32$ horseshoes
20. $\$4\times 4=\$16$
    $\$16+\$16=\$32$

Systematic Review 16F

1. $7\times 3=21$
2. $8\times 9=72$
3. $5\times 7=35$
4. $6\times 9=54$
5. $2\times 8=16$
6. $9\times 3=27$
7. $7\times 2=14$
8. $7\times 4=28$
9. $5\times 9=45$
10. $10\times 7=70$
11. $6\times 4=24$

12. $1 \times 5 = 5$
13. $2 \times \underline{8} = 16$
14. $3 \times \underline{9} = 27$
15. $16 + 16 + 16 + 16 = 64$ in
16. $8 + 8 + 8 + 8 = 32$ in
17. $5 \times 9 = 45$ sq ft
18. $5 + 9 + 5 + 9 = 28$ ft
19. $\underline{5} \times 10¢ = 50¢$; 5 dimes
20. $3 + 3 = 6$
    $6 \times 9 = 54$ buttons

Lesson Practice 17A

1. 7, 14, 21, 28, 35, 42, 49, 56, 63, 70
2. 7, 14, 21, 28, 35, 42, 49, 56, 63, 70
3. done
4. $\begin{array}{r} 20 \\ \underline{\times\ 2} \\ 40 \end{array}$
5. $\begin{array}{r} 40 \\ \underline{\times\ 3} \\ 120 \end{array}$
6. 7, 14, 21, 28, $\underline{\$35}$
7. 7, 14, $\underline{\$21}$
8. 7, 14, 21, 28, 35, $\underline{42}$ miles

Lesson Practice 17B

1. 7, 14, 21, 28, 35, 42, 49, 56, 63, 70
2. 7, 14, 21, 28, 35, 42, 49, 56, 63, 70
3. $\begin{array}{r} 40 \\ \underline{\times\ 5} \\ 200 \end{array}$
4. $\begin{array}{r} 30 \\ \underline{\times\ 6} \\ 180 \end{array}$
5. $\begin{array}{r} 20 \\ \underline{\times\ 3} \\ 60 \end{array}$
6. 7, 14, 21, 28, 35, 42, 49, $\underline{56}$ trees
7. 7, $\underline{14}$ cookies
8. 7, 14, 21, $\underline{28}$ days

Lesson Practice 17C

1. 7, 14, 21, 28, 35, 42, 49, 56, 63, 70
2. 7, 14, 21, 28, 35, 42, 49, 56, 63, 70
3. $\begin{array}{r} 20 \\ \underline{\times\ 6} \\ 120 \end{array}$
4. $\begin{array}{r} 50 \\ \underline{\times\ 3} \\ 150 \end{array}$
5. $\begin{array}{r} 80 \\ \underline{\times\ 2} \\ 160 \end{array}$
6. 7, 14, 21, 28, 35, 42, $\underline{49}$ dogs
7. 7, 14, 21, 28, 35, 42, 49, 56, $\underline{63}$ bananas
8. 7, 14, 21, 28, 35, $\underline{\$42}$

Systematic Review 17D

1. 7, 14, 21, 28, 35, 42, 49, 56, 63, 70
2. $\frac{4}{7} = \frac{8}{14} = \frac{12}{21} = \frac{16}{28} = \frac{20}{35}$
3. $4 \times 9 = 36$
4. $3 \times 8 = 24$
5. $2 \times 1 = 2$
6. $3 \times 4 = 12$
7. $7 \times 4 = 28$
8. $8 \times 6 = 48$
9. $\begin{array}{r} 40 \\ \underline{\times\ 9} \\ 360 \end{array}$
10. $\begin{array}{r} 50 \\ \underline{\times\ 2} \\ 100 \end{array}$
11. $3 \times \underline{6} = 18$
12. $8 \times \underline{10} = 80$
13. $1 \times \underline{5} = 5$
14. $7 \times \underline{9} = 63$
15. done
16. $17 = 17$
17. $31 > 13$
18. $7 \times 4 = 28$ quarters
19. $5 + 3 + 5 + 3 = 16$ in
20. $65 - 36 = 29$ pennies

Systematic Review 17E

1. 6, 12, 18, 24, 30, 36, 42, 48, 54, 60
2. $\frac{6}{7} = \frac{12}{14} = \frac{18}{21} = \frac{24}{28} = \frac{30}{35}$
3. $4 \times 6 = 24$
4. $4 \times 8 = 32$
5. $6 \times 3 = 18$
6. $9 \times 8 = 72$
7. $4 \times 4 = 16$
8. $8 \times 5 = 40$
9. $\begin{array}{r} 60 \\ \underline{\times\ 9} \\ 540 \end{array}$
10. $\begin{array}{r} 30 \\ \underline{\times\ 7} \\ 210 \end{array}$
11. $6 \times \underline{9} = 54$
12. $5 \times \underline{2} = 10$
13. $6 \times \underline{0} = 0$
14. $2 \times \underline{6} = 12$
15. $9 \text{ qt} > 8 \text{ qt}$
16. $18 = 18$
17. $15 < 18$
18. $32 + 32 + 17 = 81$ hours
19. $5 \times 10 = 50$ years
20. $2 \times 9 = 18$ pt
    $2 \times 6 = 12$ pt
    $18 + 12 = 30$ pints

Systematic Review 17F

1. 7, 14, 21, 28, 35, 42, 49, 56, 63, 70
2. $\frac{4}{5} = \frac{8}{10} = \frac{12}{15} = \frac{16}{20} = \frac{20}{25}$
3. $3 \times 4 = 12$
4. $9 \times 9 = 81$
5. $6 \times 8 = 48$
6. $4 \times 3 = 12$
7. $5 \times 9 = 45$
8. $\begin{array}{r} 10 \\ \underline{\times\ 7} \\ 70 \end{array}$
9. $\begin{array}{r} 90 \\ \underline{\times\ 3} \\ 270 \end{array}$
10. $\begin{array}{r} 60 \\ \underline{\times\ 4} \\ 240 \end{array}$
11. $7 \times \underline{2} = 14$
12. $4 \times \underline{9} = 36$
13. $6 \times \underline{7} = 42$
14. $5 \times \underline{3} = 15$
15. $36 > 30$
16. $11 < 12$
17. $18 \text{ ft} = 18 \text{ ft}$
18. $5 \times 10¢ = 50¢$
    $3 \times 5¢ = 15¢$
    $50¢ + 15¢ = 65¢$, <u>no</u>
    $75¢ - 65¢ = \underline{10¢}$
19. $8 \times 4 = 32$ lines
20. $9 \times 10 = 90$ years

Lesson Practice 18A

1. $7 \times 9 = 63$
2. $7 \times 7 = 49$
3. $10 \times 7 = 70$
4. $7 \times 6 = 42$
5. $7 \times 4 = 28$
6. $7 \times 8 = 56$
7. $7 \times 5 = 35$
8. $3 \times 7 = 21$
9. $7 \times 8 = 56$
10. $7 \times 7 = 49$
11. $7 \times 11 = 77$
    $7 \times 12 = 84$
    $7 \times 13 = 91$

12. done
13. $\begin{array}{r} 200 \\ \times\ 4 \\ \hline 800 \end{array}$
14. $\begin{array}{r} 100 \\ \times\ 9 \\ \hline 900 \end{array}$
15. $7 \times 7 = 49$ skirts
16. $6 \times 7 = 42$ days

Lesson Practice 18B

1. $7 \times 8 = 56$
2. $7 \times 4 = 28$
3. $9 \times 7 = 63$
4. $7 \times 7 = 49$
5. $7 \times 3 = 21$
6. $7 \times 5 = 35$
7. $7 \times 8 = 56$
8. $6 \times 7 = 42$
9. $\begin{array}{r} 400 \\ \times\ 2 \\ \hline 800 \end{array}$
10. $\begin{array}{r} 200 \\ \times\ 3 \\ \hline 600 \end{array}$
11. $\begin{array}{r} 100 \\ \times\ 6 \\ \hline 600 \end{array}$
12. $\begin{array}{r} 300 \\ \times\ 2 \\ \hline 600 \end{array}$
13. 7, 14, 21, 28, 35, 42, 49, 56, 63, 70
14. $\frac{0}{7 \times 0}, \frac{7}{7 \times 1}, \frac{14}{7 \times 2}, \frac{21}{7 \times 3}, \frac{28}{7 \times 4}, \frac{35}{7 \times 5},$ $\frac{42}{7 \times 6}, \frac{49}{7 \times 7}, \frac{56}{7 \times 8}, \frac{63}{7 \times 9}, \frac{70}{7 \times 10}$
15. $7 \times 7 = 49$
16. $7 \times 8 = 56$ days
17. $7 \times 2 = 14$ days
18. $7 \times 100 = 700$ soldiers

Lesson Practice 18C

1. $7 \times 7 = 49$
2. $7 \times 2 = 14$
3. $7 \times 8 = 56$
4. $9 \times 7 = 63$
5. $5 \times 7 = 35$
6. $6 \times 7 = 42$
7. $7 \times 3 = 21$
8. $7 \times 4 = 28$
9. $\begin{array}{r} 200 \\ \times\ 2 \\ \hline 400 \end{array}$
10. $\begin{array}{r} 100 \\ \times\ 5 \\ \hline 500 \end{array}$
11. $\begin{array}{r} 300 \\ \times\ 3 \\ \hline 900 \end{array}$
12. $\begin{array}{r} 200 \\ \times\ 4 \\ \hline 800 \end{array}$
13. 7, 14, 21, 28, 35, 42, 49, 56, 63, 70
14. $\frac{0}{7 \cdot 0}, \frac{7}{7 \cdot 1}, \frac{14}{7 \cdot 2}, \frac{21}{7 \cdot 3}, \frac{28}{7 \cdot 4}, \frac{35}{7 \cdot 5},$ $\frac{42}{7 \cdot 6}, \frac{49}{7 \cdot 7}, \frac{56}{7 \cdot 8}, \frac{63}{7 \cdot 9}, \frac{70}{7 \cdot 10}$
15. $7 \times 1 = 7$
16. $4 \times 7 = 28$ slices
17. $7 \times 10 = 70$ times
18. $400 \times 2 = 800$ wings

Systematic Review 18D

1. $6 \times 2 = 12$
2. $7 \times 8 = 56$
3. $3 \times 3 = 9$
4. $7 \times 2 = 14$
5. $1 \times 9 = 9$
6. $7 \times 7 = 49$
7. $10 \times 8 = 80$
8. $4 \times 3 = 12$
9. $\begin{array}{r} 70 \\ \times\ 6 \\ \hline 420 \end{array}$
10. $9 \times 7 = 63$

11. $\begin{array}{r} 40 \\ \times\ 7 \\ \hline 280 \end{array}$

12. $\begin{array}{r} 200 \\ \times\ 3 \\ \hline 600 \end{array}$

13. $3 \times \underline{9} = 27$
14. $3 \times \underline{5} = 15$
15. $20 < 21$
16. $31 > 29$
17. $35 = 35$
18. $3 \times 100 = 300$ years old
19. $A = 8 \times 7 = 56$ sq ft
    $P = 7 + 8 + 7 + 8 = 30$ ft
20. $79 + 82 + 113 = 274$ cars

Systematic Review 18E

1. $9 \times 7 = 63$
2. $2 \times 5 = 10$
3. $3 \times 6 = 18$
4. $7 \times 7 = 49$
5. $2 \times 9 = 18$
6. $8 \times 7 = 56$
7. $5 \times 6 = 30$
8. $10 \times 4 = 40$
9. $10 \times 7 = 70$
10. $6 \times 7 = 42$
11. $\begin{array}{r} 60 \\ \times\ 6 \\ \hline 360 \end{array}$

12. $\begin{array}{r} 100 \\ \times\ 8 \\ \hline 800 \end{array}$

13. $2 \times \underline{7} = 14$
14. $7 \times \underline{7} = 49$
15. $28 = 28$
16. $21 < 22$
17. $25¢ < 30¢$
18. $1 \times 100 = 100$ years old
19. $5 \times 2 = 10$
    $15 - 10 = 5$ miles
20. $5 \times 5 = 25$ sq ft
    $25 - 11 = 14$ worms

Systematic Review 18F

1. $3 \times 7 = 21$
2. $10 \times 7 = 70$
3. $3 \times 4 = 12$
4. $8 \times 7 = 56$
5. $9 \times 9 = 81$
6. $8 \times 4 = 32$
7. $7 \times 7 = 49$
8. $10 \times 2 = 20$
9. $8 \times 9 = 72$
10. $4 \times 7 = 28$
11. $\begin{array}{r} 70 \\ \times\ 5 \\ \hline 350 \end{array}$

12. $\begin{array}{r} 200 \\ \times\ 3 \\ \hline 600 \end{array}$

13. $4 \times \underline{9} = 36$
14. $4 \times \underline{6} = 24$
15. $42 > 40$
16. $360 = 360$
17. $16 = 16$
18. $9 \times 7 = 63$ temperatures
19. $3 + 5 = 8$
    $8 \times \$3 = \$24$
20. $3 + 5 = 8$
    $8 \times 5 = 40$ miles

Lesson Practice 19A

1. 8, 16, 24, 32, 40, 48, 56, 64, 72, 80
2. see #1
3. $\frac{3}{8} = \frac{6}{16} = \frac{9}{24} = \frac{12}{32} = \frac{15}{40} =$
   $\frac{18}{48} = \frac{21}{56} = \frac{24}{64} = \frac{27}{72} = \frac{30}{80}$
4. 8, 16, 24, 32, $\underline{40}$ sides
5. 8, 16, 24, 32, 40, 48, $\underline{56}$; 7 windows
6. 8, 16, 24, 32, 40, $\underline{48}$ yards
7. 8, 16, 24, 32, 40, 48, 56, $\underline{64}$ pints
8. 8, 16, 24, $\underline{32}$ legs

Lesson Practice 19B

1. 8, 16, 24, 32, 40, 48, 56, 64, 72, 80
2. see #1
3. $\frac{6}{8}=\frac{12}{16}=\frac{18}{24}=\frac{24}{32}=\frac{30}{40}=$
   $\frac{36}{48}=\frac{42}{56}=\frac{48}{64}=\frac{54}{72}=\frac{60}{80}$
4. 8, 16, 24, 32, 40, 48 pints
5. 8, 16, 24 sides
6. 8, 16, 24, 32, 40, 48, 56, $64
7. 8, 16 jugs
8. 8, 16, 24, 32, 40, 48, 56, 64, 72 legs

Lesson Practice 19C

1. 8, 16, 24, 32, 40, 48, 56, 64, 72, 80
2. see #1
3. $\frac{4}{8}=\frac{8}{16}=\frac{12}{24}=\frac{16}{32}=\frac{20}{40}=$
   $\frac{24}{48}=\frac{28}{56}=\frac{32}{64}=\frac{36}{72}=\frac{40}{80}$
4. 8, 16, 24 legs
5. 8, 16, 24, 32, 40 arms
6. $5+4=9$
   8, 16, 24, 32, 40, 48,
   56, 64, 72 arms
7. 8, 16, 24, 32, 40, 48, 56, 64,
   72, 80; 10 weeks
8. 8, 16, 24, 32, 40, 48, 56 sides

Systematic Review 19D

1. 8, 16, 24, 32, 40, 48, 56, 64, 72, 80
2. $\frac{3}{5}=\frac{6}{10}=\frac{9}{15}=\frac{12}{20}=\frac{15}{25}$
3. $8\times3=24$
4. $5\times3=15$
5. $7\times7=49$
6. $2\times4=8$
7. $7\times8=56$
8. $3\times6=18$
9. $\begin{array}{r} 40 \\ \times\ 8 \\ \hline 320 \end{array}$
10. $\begin{array}{r} 200 \\ \times\ 2 \\ \hline 400 \end{array}$
11. $9\times\underline{6}=54$
12. $4\times\underline{4}=16$
13. $8\times\underline{0}=0$
14. $7\times\underline{5}=35$
15. done
16. $\begin{array}{r} ^{1}\ \ \ \\ 128 \\ +635 \\ \hline 763 \end{array}$
17. $\begin{array}{r} 212 \\ +872 \\ \hline 1{,}084 \end{array}$
18. 4 couples = 8 people
    8, 16, 24, 32, 40, 48 dancers
19. $3\times3=9$ shoes
20. $179+143=322$ laps

Systematic Review 19E

1. 7, 14, 21, 28, 35, 42, 49, 56, 63, 70
2. $\frac{2}{6}=\frac{4}{12}=\frac{6}{18}=\frac{8}{24}=\frac{10}{30}$
3. $8\times2=16$
4. $5\times8=40$
5. $9\times8=72$
6. $3\times6=18$
7. $7\times7=49$
8. $10\times5=50$
9. $\begin{array}{r} 60 \\ \times\ 7 \\ \hline 420 \end{array}$
10. $\begin{array}{r} 100 \\ \times\ 9 \\ \hline 900 \end{array}$
11. $3\times\underline{9}=27$
12. $6\times\underline{6}=36$
13. $5\times\underline{1}=5$
14. $4\times\underline{5}=20$
15. $\begin{array}{r} 601 \\ +513 \\ \hline 1{,}114 \end{array}$

16. $\begin{array}{r} {\scriptstyle 11} \\ 245 \\ +189 \\ \hline 434 \end{array}$

17. $\begin{array}{r} 538 \\ +251 \\ \hline 789 \end{array}$

18. $3+4=7$ gallons
    8, 16, 24, 32, 40, 48, <u>56</u> pints

19. $2\times4=8$
    $8+2=10$ lb

20. $74+98+206=378$ coins

Systematic Review 19F

1. 8, 16, 24, 32, 40, 48, 56, 64, 72, 80
2. $\frac{1}{7}=\frac{2}{14}=\frac{3}{21}=\frac{4}{28}=\frac{5}{35}$
3. $8\times10=80$
4. $6\times8=48$
5. $8\times4=32$
6. $3\times10=30$
7. $1\times1=1$
8. $\begin{array}{r} 80 \\ \times\ 4 \\ \hline 320 \end{array}$
9. $\begin{array}{r} 10 \\ \times\ 5 \\ \hline 50 \end{array}$
10. $\begin{array}{r} 100 \\ \times\ 2 \\ \hline 200 \end{array}$
11. $8\times\underline{3}=24$
12. $7\times\underline{7}=49$
13. $4\times\underline{3}=12$
14. $7\times\underline{2}=14$
15. $\begin{array}{r} {\scriptstyle 1} \\ 452 \\ +318 \\ \hline 770 \end{array}$
16. $\begin{array}{r} 711 \\ +206 \\ \hline 917 \end{array}$
17. $\begin{array}{r} {\scriptstyle 1} \\ 153 \\ +592 \\ \hline 745 \end{array}$
18. $\$45-\$27=\$18$
19. $6\times5=30$
    $30-7=23$ gallons
20. $5\times10=50$ sq ft
    $50-35=15$ tiles

Lesson Practice 20A

1. $8\times9=72$
2. $7\times8=56$
3. $6\times8=48$
4. $8\times10=80$
5. $5\times8=40$
6. $1\times8=8$
7. $8\times3=24$
8. $8\times2=16$
9. $8\times8=64$
10. $8\times9=72$
11. $4\times8=32$
12. $8\times7=56$
13. $8\times0=0$
14. $8\times8=64$
15.

								■	
						■			
				■					
		■							
■								■	
						■			
				■					
		■							
■								■	
						■			

16. $8\times10=80$ sides
17. $6\times8=48$ sides
18. $3\times8=24$ oranges

Lesson Practice 20B

1. $8 \times 4 = 32$
2. $8 \times 8 = 64$
3. $5 \times 8 = 40$
4. $7 \times 8 = 56$
5. $9 \times 8 = 72$
6. $10 \times 8 = 80$
7. $8 \times 6 = 48$
8. $8 \times 3 = 24$
9. $9 \times 8 = 72$
10. $8 \times 2 = 16$
11. $8 \times 8 = 64$
12. $8 \times 1 = 8$
13. 8, 16, 24, 32, 40, 48, 56, 64, 72, 80
14. $\frac{0}{(8)(0)}, \frac{8}{(8)(1)}, \frac{16}{(8)(2)}, \frac{24}{(8)(3)},$
$\frac{32}{(8)(4)}, \frac{40}{(8)(5)}, \frac{48}{(8)(6)}, \frac{56}{(8)(7)},$
$\frac{64}{(8)(8)}, \frac{72}{(8)(9)}, \frac{80}{(8)(10)}$
15. $5 \times 8 = 40$ sides
16. $8 \times 4 = 32$
17. $8 \times 7 = 56$ slices
18. $9 \times 8 = 72$ people

Lesson Practice 20C

1. $6 \times 8 = 48$
2. $8 \times 2 = 16$
3. $8 \times 8 = 64$
4. $8 \times 9 = 72$
5. $8 \times 3 = 24$
6. $5 \times 8 = 40$
7. $7 \times 8 = 56$
8. $8 \times 10 = 80$
9. $4 \times 8 = 32$
10. $8 \times 1 = 8$
11. $6 \times 8 = 48$
12. $8 \times 9 = 72$
13. $8 \times 8 = 64$
14. $8 \times 5 = 40$
15. 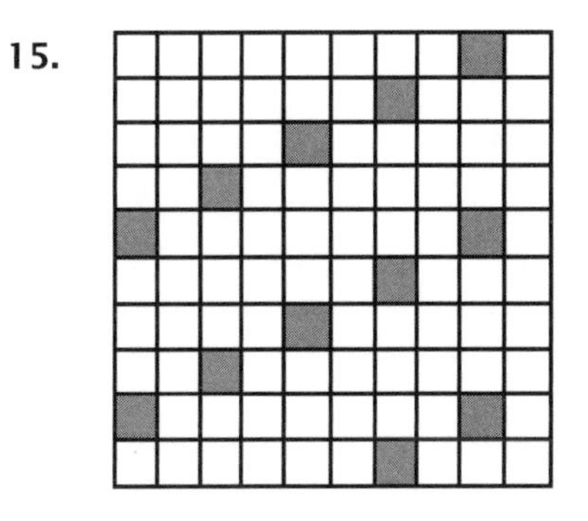
$8 \times 11 = 88$
$8 \times 12 = 96$
16. $8 \times 7 = 56$ sides
17. $8 \times 4 = 32$ sides
18. $8 \times 8 = 64$ cylinders

Systematic Review 20D

1. $7 \times 7 = 49$
2. $8 \times 6 = 48$
3. $9 \times 9 = 81$
4. $3 \times 1 = 3$
5. $8 \times 7 = 56$
6. $9 \times 3 = 27$
7. $7 \times 6 = 42$
8. $9 \times 5 = 45$
9. $\begin{array}{r} 80 \\ \times\ 8 \\ \hline 640 \end{array}$
10. $2 \times 0 = 0$
11. $\begin{array}{r} 40 \\ \times\ 4 \\ \hline 160 \end{array}$
12. $\begin{array}{r} 100 \\ \times\ 3 \\ \hline 300 \end{array}$
13. $49 > 14$
14. $41 = 41$
15. $24 < 28$
16. done
17. done

18. $$\begin{array}{r} \overset{4}{\cancel{5}}\,\overset{15}{\cancel{6}}\,{}^{1}2 \\ -\,3\,7\,4 \\ \hline 1\,8\,8 \end{array}$$

19. $3+4=7$
$7\times8=56$ legs

20. $458-328=130$

Systematic Review 20E

1. $8\times8=64$
2. $7\times7=49$
3. $2\times6=12$
4. $5\times4=20$
5. $9\times4=36$
6. $6\times6=36$
7. $7\times8=56$
8. $3\times7=21$
9. $$\begin{array}{r} 40 \\ \times\ \ 3 \\ \hline 120 \end{array}$$
10. $2\times8=16$
11. $$\begin{array}{r} 80 \\ \times\ \ 3 \\ \hline 240 \end{array}$$
12. $$\begin{array}{r} 200 \\ \times\ \ 2 \\ \hline 400 \end{array}$$
13. $45¢>40¢$
14. $24=24$
15. $32<33$
16. $$\begin{array}{r} \overset{5}{\cancel{6}}\,\overset{9}{{}^{1}\cancel{0}}\,{}^{1}3 \\ -\,1\,1\,8 \\ \hline 4\,8\,5 \end{array}$$
17. $$\begin{array}{r} 345 \\ -142 \\ \hline 203 \end{array}$$
18. $$\begin{array}{r} 8\,\overset{2}{\cancel{3}}\,{}^{1}7 \\ -1\,0\,8 \\ \hline 7\,2\,9 \end{array}$$

19. $8\times4=32$
$8\times2=16$
$16+32=48$ tires

20. $244-188=56$ pencils

Systematic Review 20F

1. $3\times6=18$
2. $9\times7=63$
3. $8\times4=32$
4. $3\times9=27$
5. $8\times8=64$
6. $5\times0=0$
7. $3\times3=9$
8. $7\times7=49$
9. $$\begin{array}{r} 70 \\ \times\ \ 4 \\ \hline 280 \end{array}$$
10. $9\times9=81$
11. $$\begin{array}{r} 60 \\ \times\ \ 3 \\ \hline 180 \end{array}$$
12. $$\begin{array}{r} 400 \\ \times\ \ 2 \\ \hline 800 \end{array}$$
13. $0<9$
14. $47>18$
15. 16 qt <20 qt
16. $$\begin{array}{r} \overset{8}{\cancel{9}}\,\overset{9}{{}^{1}\cancel{0}}\,{}^{1}0 \\ -\,1\,2\,3 \\ \hline 7\,7\,7 \end{array}$$
17. $$\begin{array}{r} 6\,\overset{7}{\cancel{8}}\,{}^{1}3 \\ -2\,5\,4 \\ \hline 4\,2\,9 \end{array}$$
18. $$\begin{array}{r} \overset{4}{\cancel{5}}\,{}^{1}0\,6 \\ -\,3\,4\,4 \\ \hline 1\,6\,2 \end{array}$$
19. $549+86=635$
$635-350=\$285$
20. $2\times\underline{9}=18$; 9 geese

Lesson Practice 21A

1. done

2. $\begin{array}{r} 13 \\ \times\ 2 \\ \hline 26 \end{array} \qquad \begin{array}{r} 10+3 \\ \times\quad 2 \\ \hline 20+6 \end{array}$

3. $\begin{array}{r} 11 \\ \times\ 7 \\ \hline 77 \end{array} \qquad \begin{array}{r} 10+1 \\ \times\quad 7 \\ \hline 70+7 \end{array}$

4. $\begin{array}{r} 21 \\ \times\ 2 \\ \hline 42 \end{array} \qquad \begin{array}{r} 20+1 \\ \times\quad 2 \\ \hline 40+2 \end{array}$

5. $\begin{array}{r} 32 \\ \times\ 3 \\ \hline 96 \end{array} \qquad \begin{array}{r} 30+2 \\ \times\quad 3 \\ \hline 90+6 \end{array}$

6. $\begin{array}{r} 14 \\ \times\ 2 \\ \hline 28 \end{array} \qquad \begin{array}{r} 10+4 \\ \times\quad 2 \\ \hline 20+8 \end{array}$

7. $\begin{array}{r} 11 \\ \times\ 9 \\ \hline 99 \end{array} \qquad \begin{array}{r} 10+1 \\ \times\quad 9 \\ \hline 90+9 \end{array}$

8. $\begin{array}{r} 24 \\ \times\ 2 \\ \hline 48 \end{array} \qquad \begin{array}{r} 20+4 \\ \times\quad 2 \\ \hline 40+8 \end{array}$

9. $\begin{array}{r} 123 \\ \times\ 2 \\ \hline 246 \end{array} \qquad \begin{array}{r} 100+20+3 \\ \times\qquad 2 \\ \hline 200+40+6 \end{array}$

10. $\begin{array}{r} 222 \\ \times\ 4 \\ \hline 888 \end{array} \qquad \begin{array}{r} 200+20+2 \\ \times\qquad 4 \\ \hline 800+80+8 \end{array}$

11. $4 \times 12 = 48\ (40+8)$ books

12. $310 \times 3 = 930\ (900+30)$ shots

Lesson Practice 21B

1. $\begin{array}{r} 21 \\ \times\ 3 \\ \hline 63 \end{array} \qquad \begin{array}{r} 20+1 \\ \times\quad 3 \\ \hline 60+3 \end{array}$

2. $\begin{array}{r} 24 \\ \times\ 2 \\ \hline 48 \end{array} \qquad \begin{array}{r} 20+4 \\ \times\quad 2 \\ \hline 40+8 \end{array}$

3. $\begin{array}{r} 22 \\ \times\ 4 \\ \hline 88 \end{array} \qquad \begin{array}{r} 20+2 \\ \times\quad 4 \\ \hline 80+8 \end{array}$

4. $\begin{array}{r} 11 \\ \times\ 6 \\ \hline 66 \end{array} \qquad \begin{array}{r} 10+1 \\ \times\quad 6 \\ \hline 60+6 \end{array}$

5. $\begin{array}{r} 14 \\ \times\ 2 \\ \hline 28 \end{array} \qquad \begin{array}{r} 10+4 \\ \times\quad 2 \\ \hline 20+8 \end{array}$

6. $\begin{array}{r} 33 \\ \times\ 3 \\ \hline 99 \end{array} \qquad \begin{array}{r} 30+3 \\ \times\quad 3 \\ \hline 90+9 \end{array}$

7. $\begin{array}{r} 110 \\ \times\ 5 \\ \hline 550 \end{array} \qquad \begin{array}{r} 100+10+0 \\ \times\qquad 5 \\ \hline 500+50+0 \end{array}$

8. $\begin{array}{r} 231 \\ \times\ 3 \\ \hline 693 \end{array} \qquad \begin{array}{r} 200+30+1 \\ \times\qquad 3 \\ \hline 600+90+3 \end{array}$

9. $\begin{array}{r} 424 \\ \times\ 2 \\ \hline 848 \end{array} \qquad \begin{array}{r} 400+20+4 \\ \times\qquad 2 \\ \hline 800+40+8 \end{array}$

10. $\begin{array}{r} 121 \\ \times\ 4 \\ \hline 484 \end{array} \qquad \begin{array}{r} 100+20+1 \\ \times\qquad 4 \\ \hline 400+80+4 \end{array}$

11. $2 \times 12 = 24$ sq in

12. $2 \times 214 = 428$ legs

Lesson Practice 21C

1. $\begin{array}{r} 43 \\ \times\ 2 \\ \hline 86 \end{array} \qquad \begin{array}{r} 40+3 \\ \times\quad 2 \\ \hline 80+6 \end{array}$

2. $\begin{array}{r} 32 \\ \times\ 2 \\ \hline 64 \end{array} \qquad \begin{array}{r} 30+2 \\ \times\quad 2 \\ \hline 60+4 \end{array}$

3. $\begin{array}{r} 12 \\ \times\ 3 \\ \hline 36 \end{array} \qquad \begin{array}{r} 10+2 \\ \times\quad 3 \\ \hline 30+6 \end{array}$

4. $\begin{array}{r} 11 \\ \times\ 4 \\ \hline 44 \end{array} \qquad \begin{array}{r} 10+1 \\ \times\quad 4 \\ \hline 40+4 \end{array}$

5. $\begin{array}{r} 42 \\ \times\ 2 \\ \hline 84 \end{array} \qquad \begin{array}{r} 40+2 \\ \times\quad 2 \\ \hline 80+4 \end{array}$

6. $\begin{array}{r} 31 \\ \times\ 3 \\ \hline 93 \end{array} \qquad \begin{array}{r} 30+1 \\ \times\quad 3 \\ \hline 90+3 \end{array}$

7. $\begin{array}{r} 413 \\ \times\ 2 \\ \hline 826 \end{array}$ $\begin{array}{r} 400+10+3 \\ \times\ \ \ \ \ 2 \\ \hline 800+20+6 \end{array}$

8. $\begin{array}{r} 111 \\ \times\ 6 \\ \hline 666 \end{array}$ $\begin{array}{r} 100+10+1 \\ \times\ \ \ \ \ 6 \\ \hline 600+60+6 \end{array}$

9. $\begin{array}{r} 103 \\ \times\ 3 \\ \hline 309 \end{array}$ $\begin{array}{r} 100+0+3 \\ \times\ \ \ \ \ 3 \\ \hline 300+0+9 \end{array}$

10. $\begin{array}{r} 212 \\ \times\ 4 \\ \hline 848 \end{array}$ $\begin{array}{r} 200+10+2 \\ \times\ \ \ \ \ 4 \\ \hline 800+40+8 \end{array}$

11. $4 \times 21 = 84$ plants

12. $8 \times 111 = 888$ legs

Systematic Review 21D

1. $\begin{array}{r} 11 \\ \times\ 5 \\ \hline 55 \end{array}$ $\begin{array}{r} 10+1 \\ \times\ \ \ 5 \\ \hline 50+5 \end{array}$

2. $\begin{array}{r} 12 \\ \times\ 3 \\ \hline 36 \end{array}$ $\begin{array}{r} 10+2 \\ \times\ \ \ 3 \\ \hline 30+6 \end{array}$

3. $\begin{array}{r} 324 \\ \times\ 2 \\ \hline 648 \end{array}$ $\begin{array}{r} 300+20+4 \\ \times\ \ \ \ \ 2 \\ \hline 600+40+8 \end{array}$

4. $\begin{array}{r} 322 \\ \times\ 3 \\ \hline 966 \end{array}$ $\begin{array}{r} 300+20+2 \\ \times\ \ \ \ \ 3 \\ \hline 900+60+6 \end{array}$

5. $8 \times 7 = 56$
6. $8 \times 8 = 64$
7. $7 \times 7 = 49$
8. $4 \times 6 = 24$
9. done
10. +
11. –
12. ×

13. $\begin{array}{r} {}^{1}\ \ \\ 23 \\ 45 \\ +17 \\ \hline 85 \end{array}$

14. $\begin{array}{r} {}^{2}\ \ \\ 39 \\ 24 \\ +88 \\ \hline 151 \end{array}$

15. $\begin{array}{r} {}^{4}\ \ \ \\ 4\not{5}{}^{1}2 \\ -129 \\ \hline 323 \end{array}$

16. $\begin{array}{r} {}^{7}\ \ \ \\ 2\not{8}{}^{1}3 \\ -216 \\ \hline 67 \end{array}$

17. $p = 8+11+8+11 = 38$ ft
$a = 8 \times 11 = 88$ sq ft

18. $9 \times 7 = 63$
$63+12 = 75$ discoveries

19. $18+52 = 70$ pages

20. $150-123 = 27$ logs

Systematic Review 21E

1. $\begin{array}{r} 12 \\ \times\ 4 \\ \hline 48 \end{array}$ $\begin{array}{r} 10+2 \\ \times\ \ \ 4 \\ \hline 40+8 \end{array}$

2. $\begin{array}{r} 32 \\ \times\ 3 \\ \hline 96 \end{array}$ $\begin{array}{r} 30+2 \\ \times\ \ \ 3 \\ \hline 90+6 \end{array}$

3. $\begin{array}{r} 221 \\ \times\ 4 \\ \hline 884 \end{array}$ $\begin{array}{r} 200+20+1 \\ \times\ \ \ \ \ 4 \\ \hline 800+80+4 \end{array}$

4. $\begin{array}{r} 313 \\ \times\ 2 \\ \hline 626 \end{array}$ $\begin{array}{r} 300+10+3 \\ \times\ \ \ \ \ 2 \\ \hline 600+20+6 \end{array}$

5. $8 \times 4 = 32$
6. $6 \times 8 = 48$
7. $7 \times 9 = 63$
8. $9 \times 4 = 36$
9. +
10. ×
11. ×
12. –

13. $$\begin{array}{r} 91 \\ 25 \\ +42 \\ \hline 158 \end{array}$$

14. $$\begin{array}{r} {}^{1} \\ 67 \\ 13 \\ +50 \\ \hline 130 \end{array}$$

15. $$\begin{array}{r} 8\overset{8}{\not 9}{}^{1}3 \\ -615 \\ \hline 278 \end{array}$$

16. $$\begin{array}{r} \overset{2}{\not 3}{}^{1}14 \\ -\ 92 \\ \hline 222 \end{array}$$

17. $p = 6+7+6+7 = 26$ ft
    $a = 6 \times 7 = 42$ sq ft
18. $200 \times 3 = 600$ fish
19. $3 \times 12 = 36$ doughnuts
    $36 + 36 = 72$ doughnuts
20. $17+22+24+31 = 94$ pushups

Systematic Review 21F

1. $$\begin{array}{r} 21 \\ \times\ 2 \\ \hline 42 \end{array} \qquad \begin{array}{r} 20+1 \\ \times\ \ \ 2 \\ \hline 40+2 \end{array}$$

2. $$\begin{array}{r} 11 \\ \times\ 6 \\ \hline 66 \end{array} \qquad \begin{array}{r} 10+1 \\ \times\ \ \ 6 \\ \hline 60+6 \end{array}$$

3. $$\begin{array}{r} 202 \\ \times\ 3 \\ \hline 606 \end{array} \qquad \begin{array}{r} 200+0+2 \\ \times\ \ \ 3 \\ \hline 600+0+6 \end{array}$$

4. $$\begin{array}{r} 444 \\ \times\ 2 \\ \hline 888 \end{array} \qquad \begin{array}{r} 400+40+4 \\ \times\ \ \ 2 \\ \hline 800+80+8 \end{array}$$

5. $8 \times 6 = 48$
6. $3 \times 7 = 21$
7. $6 \times 10 = 60$
8. $9 \times 7 = 63$
9. $-$
10. $\times$
11. $+$
12. $\times$

13. $$\begin{array}{r} {}^{1} \\ 82 \\ 13 \\ +56 \\ \hline 151 \end{array}$$

14. $$\begin{array}{r} {}^{1} \\ 71 \\ 64 \\ +26 \\ \hline 161 \end{array}$$

15. $$\begin{array}{r} 5\overset{5}{\not 6}{}^{1}1 \\ -\ 19 \\ \hline 542 \end{array}$$

16. $$\begin{array}{r} \overset{3}{\not 4}{}^{1}37 \\ -365 \\ \hline 72 \end{array}$$

17. $15+12+15+12 = 54$ ft
18. $2 \times 431 = 862$ birds
19. $7+2 = 9$ fish per hour
    $9 \times 6 = 54$ fish
20. $7 \times 11 = 77$ eggs
    $77 - 18 = 59$ left over

Lesson Practice 22A

1. 40
2. 10
3. 60
4. 60
5. 40
6. 60
7. 400
8. 200
9. 700
10. 2,000
11. 2,000
12. 7,000
13. done
14. $30 \times 6 = 180$
15. $20 \times 3 = 60$
16. done
17. $400 \times 3 = 1{,}200$

18. $100 \times 4 = 400$
19. $30 \times 5 = 150$ cars
20. $40 \times 3 = \$120$

Lesson Practice 22B

1. 30
2. 70
3. 50
4. 500
5. 700
6. 200
7. 900
8. 100
9. 200
10. 6,000
11. 9,000
12. 4,000
13. $10 \times 2 = 20$
14. $30 \times 2 = 60$
15. $20 \times 7 = 140$
16. $400 \times 3 = 1{,}200$
17. $100 \times 4 = 400$
18. $600 \times 2 = 1{,}200$
19. $20 \times 5 = 100$ mi
20. $50 \times 3 = \$150$

Lesson Practice 22C

1. 90
2. 40
3. 20
4. 200
5. 500
6. 900
7. 400
8. 200
9. 600
10. 8,000
11. 1,000
12. 3,000
13. $50 \times 2 = 100$
14. $40 \times 3 = 120$
15. $20 \times 5 = 100$
16. $100 \times 8 = 800$
17. $300 \times 2 = 600$
18. $400 \times 6 = 2{,}400$
19. $70 \times 9 = 630$ books
20. $20 \times 4 = 80$ people

Systematic Review 22D

1. 70
2. 10
3. 50
4. $10 \times 8 = 80$
5. $80 \times 9 = 720$
6. $60 \times 5 = 300$
7. $500 \times 3 = 1{,}500$
8. $400 \times 6 = 2{,}400$
9. $900 \times 2 = 1{,}800$
10. $\begin{array}{r} 11 \\ \times\ 6 \\ \hline 66 \end{array} \quad \begin{array}{r} 10+1 \\ \times\ \ \ 6 \\ \hline 60+6 \end{array}$
11. $\begin{array}{r} 21 \\ \times\ 4 \\ \hline 84 \end{array} \quad \begin{array}{r} 20+1 \\ \times\ \ \ 4 \\ \hline 80+4 \end{array}$
12. $\begin{array}{r} 113 \\ \times\ 3 \\ \hline 339 \end{array} \quad \begin{array}{r} 100+10+3 \\ \times\ \ \ 3 \\ \hline 300+30+9 \end{array}$
13. $\begin{array}{r} 423 \\ \times\ 2 \\ \hline 846 \end{array} \quad \begin{array}{r} 400+20+3 \\ \times\ \ \ 2 \\ \hline 800+40+6 \end{array}$
14. $56 > 54$
15. 6 qt $<$ 8 qt
16. $28 > 24$
17. $200 \times 3 = \$600$
18. $53 \times 2 = 106$ jars
19. $4 \times 12 = 48$ soldiers
20. $2 \times 7 = 14$; $3 \times 3 = 9$
    $14 + 9 = 23$ pts. (our team)
    $1 \times 7 = 7$; $6 \times 3 = 18$
    $7 + 18 = 25$ pts. (other team)
    $23 < 25$, so the other team won.

Systematic Review 22E

1. 400
2. 200
3. 600
4. $20 \times 4 = 80$
5. $70 \times 5 = 350$
6. $90 \times 6 = 540$
7. $200 \times 8 = 1{,}600$
8. $500 \times 4 = 2{,}000$
9. $400 \times 3 = 1{,}200$
10. $$\begin{array}{r} 44 \\ \times\ 2 \\ \hline 88 \end{array} \qquad \begin{array}{r} 40+4 \\ \times\qquad 2 \\ \hline 80+8 \end{array}$$
11. $$\begin{array}{r} 32 \\ \times\ 2 \\ \hline 64 \end{array} \qquad \begin{array}{r} 30+2 \\ \times\qquad 2 \\ \hline 60+4 \end{array}$$
12. $$\begin{array}{r} 303 \\ \times\ 3 \\ \hline 909 \end{array} \qquad \begin{array}{r} 300+0+3 \\ \times\qquad 3 \\ \hline 900+0+9 \end{array}$$
13. $$\begin{array}{r} 122 \\ \times\ 4 \\ \hline 488 \end{array} \qquad \begin{array}{r} 100+20+2 \\ \times\qquad 4 \\ \hline 400+80+8 \end{array}$$
14. $64 > 60$
15. $40¢ = 40¢$
16. 40 pt < 45 pt
17. $2 \times 12 = 24$ pints
18. $2 \times 23 = 46$
    $46 + 24 = 70$ pints
19. $13 + 12 + 13 + 12 = 50$
    $50 = 50$; yes
20. $\$50 - \$19 = \$31$

Systematic Review 22F

1. 3,000
2. 4,000
3. 8,000
4. $80 \times 7 = 560$
5. $60 \times 3 = 180$
6. $50 \times 4 = 200$
7. $900 \times 3 = 2{,}700$
8. $700 \times 2 = 1{,}400$
9. $300 \times 5 = 1{,}500$
10. $$\begin{array}{r} 13 \\ \times\ 3 \\ \hline 39 \end{array} \qquad \begin{array}{r} 10+3 \\ \times\qquad 3 \\ \hline 30+9 \end{array}$$
11. $$\begin{array}{r} 41 \\ \times\ 2 \\ \hline 82 \end{array} \qquad \begin{array}{r} 40+1 \\ \times\qquad 2 \\ \hline 80+2 \end{array}$$
12. $$\begin{array}{r} 111 \\ \times\ 8 \\ \hline 888 \end{array} \qquad \begin{array}{r} 100+10+1 \\ \times\qquad 8 \\ \hline 800+80+8 \end{array}$$
13. $$\begin{array}{r} 323 \\ \times\ 3 \\ \hline 969 \end{array} \qquad \begin{array}{r} 300+20+3 \\ \times\qquad 3 \\ \hline 900+60+9 \end{array}$$
14. $24 = 24$
15. $56¢ < 60¢$
16. $36 > 32$
17. $24 + 4 + 3 + 1 = 32$ animals
18. $\$2 \times 24 = \$48$
    $\$12 \times 4 = \$48$
    $\$6 \times 3 = \$18$
    $\$28 \times 1 = \$28$
    $48 + 48 + 18 + 28 = \$142$
19. $70 \times 6 = 420$ gallons
20. $50 \times 7 = 350$ gallons
    $420 > 350$

Lesson Practice 23A

1. done
2. $11 \times 11 = 121$
3. done
4. $$\begin{array}{r} 32 \\ \times 11 \\ \hline 32 \\ 32 \\ \hline 352 \end{array} \qquad \begin{array}{r} 30+2 \\ \hline \times 10+1 \\ \hline 30+2 \\ 300+20 \\ \hline 300+50+2 \end{array}$$
5. $$\begin{array}{r} 22 \\ \times 10 \\ \hline 0 \\ 22 \\ \hline 220 \end{array} \qquad \begin{array}{r} 20+2 \\ \hline \times 10+0 \\ \hline +0 \\ 200+20 \\ \hline 200+20+0 \end{array}$$
6. $$\begin{array}{r} 23 \\ \times 13 \\ \hline 69 \\ 23 \\ \hline 299 \end{array} \qquad \begin{array}{r} 20+3 \\ \hline \times 10+3 \\ \hline 60+9 \\ 200+30 \\ \hline 200+90+9 \end{array}$$

7. $$\begin{array}{r} 12 \\ \times 12 \\ \hline 24 \\ 12 \\ \hline 144 \end{array} \qquad \begin{array}{r} 10+2 \\ \times 10+2 \\ \hline 20+4 \\ 100+20 \\ \hline 100+40+4 \end{array}$$

8. $$\begin{array}{r} 21 \\ \times 14 \\ \hline 84 \\ 21 \\ \hline 294 \end{array} \qquad \begin{array}{r} 20+1 \\ \times 10+4 \\ \hline 80+4 \\ 200+10 \\ \hline 200+90+4 \end{array}$$

9. $\$12 \times 13 = \$156$

10. $17 \times 11 = 187$ pets

Lesson Practice 23B

1. $12 \times 22 = 264$
2. $12 \times 13 = 156$

3. $$\begin{array}{r} 13 \\ \times 11 \\ \hline 13 \\ 13 \\ \hline 143 \end{array} \qquad \begin{array}{r} 10+3 \\ \times 10+1 \\ \hline 10+3 \\ 100+30 \\ \hline 100+40+3 \end{array}$$

4. $$\begin{array}{r} 21 \\ \times 12 \\ \hline 42 \\ 21 \\ \hline 252 \end{array} \qquad \begin{array}{r} 20+1 \\ \times 10+2 \\ \hline 40+2 \\ 200+10 \\ \hline 200+50+2 \end{array}$$

5. $$\begin{array}{r} 45 \\ \times 11 \\ \hline 45 \\ 45 \\ \hline 495 \end{array} \qquad \begin{array}{r} 40+5 \\ \times 10+1 \\ \hline 40+5 \\ 400+50 \\ \hline 400+90+5 \end{array}$$

6. $$\begin{array}{r} 23 \\ \times 21 \\ \hline 23 \\ 46 \\ \hline 483 \end{array} \qquad \begin{array}{r} 20+3 \\ \times 20+1 \\ \hline 20+3 \\ 400+60 \\ \hline 400+80+3 \end{array}$$

7. $$\begin{array}{r} 22 \\ \times 13 \\ \hline 66 \\ 22 \\ \hline 286 \end{array} \qquad \begin{array}{r} 20+2 \\ \times 10+3 \\ \hline 60+6 \\ 200+20 \\ \hline 200+80+6 \end{array}$$

8. $$\begin{array}{r} 37 \\ \times 10 \\ \hline 0 \\ 37 \\ \hline 370 \end{array} \qquad \begin{array}{r} 30+7 \\ \times 10+0 \\ \hline +0 \\ 300+70 \\ \hline 300+70+0 \end{array}$$

9. $20 \times 12 = 240$ months

10. $11 \times 41 = 451$ plants

Lesson Practice 23C

1. $21 \times 11 = 231$
2. $11 \times 12 = 132$

3. $$\begin{array}{r} 22 \\ \times 11 \\ \hline 22 \\ 22 \\ \hline 242 \end{array} \qquad \begin{array}{r} 20+2 \\ \times 10+1 \\ \hline 20+2 \\ 200+20 \\ \hline 200+40+2 \end{array}$$

4. $$\begin{array}{r} 33 \\ \times 12 \\ \hline 66 \\ 33 \\ \hline 396 \end{array} \qquad \begin{array}{r} 30+3 \\ \times 10+2 \\ \hline 60+6 \\ 300+30 \\ \hline 300+90+6 \end{array}$$

5. $$\begin{array}{r} 11 \\ \times 16 \\ \hline 66 \\ 11 \\ \hline 176 \end{array} \qquad \begin{array}{r} 10+1 \\ \times 10+6 \\ \hline 60+6 \\ 100+10 \\ \hline 100+70+6 \end{array}$$

6. $$\begin{array}{r} 19 \\ \times 10 \\ \hline 0 \\ 19 \\ \hline 190 \end{array} \qquad \begin{array}{r} 10+9 \\ \times 10+0 \\ \hline +0 \\ 100+90 \\ \hline 100+90+0 \end{array}$$

7. $$\begin{array}{r} 22 \\ \times 22 \\ \hline 44 \\ 44 \\ \hline 484 \end{array} \qquad \begin{array}{r} 20+2 \\ \times 20+2 \\ \hline 40+4 \\ 400+40 \\ \hline 400+80+4 \end{array}$$

8. $$\begin{array}{r} 44 \\ \times 11 \\ \hline 44 \\ 44 \\ \hline 484 \end{array} \qquad \begin{array}{r} 40+4 \\ \times 10+1 \\ \hline 40+4 \\ 400+40 \\ \hline 400+80+4 \end{array}$$

9. $15 \times 11 = 165$ birds

10. $13 \times 11 = 143$ sq ft

Systematic Review 23D

1. $$\begin{array}{r} 20 \\ \underline{\times 13} \\ 60 \\ \underline{20} \\ 260 \end{array}$$

2. $$\begin{array}{r} 27 \\ \underline{\times 11} \\ 27 \\ \underline{27} \\ 297 \end{array}$$

3. $$\begin{array}{r} 13 \\ \underline{\times 13} \\ 39 \\ \underline{13} \\ 169 \end{array}$$

4. $$\begin{array}{r} 24 \\ \underline{\times 21} \\ 24 \\ \underline{\overset{1}{4}8} \\ 504 \end{array}$$

5. $$\begin{array}{r} 12 \\ \underline{\times 31} \\ 12 \\ \underline{36} \\ 372 \end{array}$$

6. $$\begin{array}{r} 23 \\ \underline{\times 12} \\ 46 \\ \underline{23} \\ 276 \end{array}$$

7. $$\begin{array}{r} 12 \\ \underline{\times 44} \\ 48 \\ \underline{\overset{1}{4}8} \\ 528 \end{array}$$

8. $$\begin{array}{r} 31 \\ \underline{\times 21} \\ 31 \\ \underline{62} \\ 651 \end{array}$$

9. 600
10. 300
11. 800
12. $20 \times 9 = 180$
13. $30 \times 7 = 210$
14. $30 \times 6 = 180$
15. $$\begin{array}{r} \overset{4}{\cancel{5}}\,{}^{1}19 \\ \underline{-\,124} \\ 395 \end{array}$$

16. $$\begin{array}{r} \overset{7}{\cancel{8}}\,{}^{1}63 \\ \underline{-\,293} \\ 570 \end{array}$$

17. $$\begin{array}{r} \overset{5}{\cancel{6}}\,\overset{{}^{1}0}{\cancel{1}}\,{}^{1}1 \\ \underline{-\,299} \\ 312 \end{array}$$

18. $113 \times 2 = 226$ ants
19. $212 \times 4 = 848$ salmon
20. $100 \times 2 = 200$ pints

Systematic Review 23E

1. $$\begin{array}{r} 30 \\ \underline{\times 13} \\ 90 \\ \underline{30} \\ 390 \end{array}$$

2. $$\begin{array}{r} 21 \\ \underline{\times 20} \\ 0 \\ \underline{42} \\ 420 \end{array}$$

3. $$\begin{array}{r} 31 \\ \underline{\times 12} \\ 62 \\ \underline{31} \\ 372 \end{array}$$

4. $$\begin{array}{r} 44 \\ \underline{\times 22} \\ 88 \\ \underline{\overset{1}{8}8} \\ 968 \end{array}$$

5. $\begin{array}{r} 22 \\ \times 33 \\ \hline 66 \\ \overset{1}{66} \\ \hline 726 \end{array}$

6. $\begin{array}{r} 13 \\ \times 11 \\ \hline 13 \\ 13 \\ \hline 143 \end{array}$

7. $\begin{array}{r} 21 \\ \times 22 \\ \hline 42 \\ 42 \\ \hline 462 \end{array}$

8. $\begin{array}{r} 10 \\ \times 21 \\ \hline 10 \\ 20 \\ \hline 210 \end{array}$

9. 90

10. 40

11. 30

12. $700 \times 3 = 2{,}100$

13. $100 \times 5 = 500$

14. $300 \times 8 = 2{,}400$

15. $\begin{array}{r} \overset{0}{\cancel{1}}\,{}^{1}16 \\ -\ 95 \\ \hline 21 \end{array}$

16. $\begin{array}{r} \overset{6}{\cancel{7}}\,{}^{1}\overset{9}{\cancel{0}}\,{}^{1}0 \\ -\ 6\ 0\ 2 \\ \hline 9\ 8 \end{array}$

17. $\begin{array}{r} 5\overset{4}{\cancel{5}}\,{}^{1}8 \\ -349 \\ \hline 209 \end{array}$

18. $44 + 26 = 70$
$70 - 21 = 49$ tons

19. $40 + 50 = \$90$

20. $\$60 \times 9 = \$540$

Systematic Review 23F

1. $\begin{array}{r} 35 \\ \times 11 \\ \hline 35 \\ 35 \\ \hline 385 \end{array}$

2. $\begin{array}{r} 23 \\ \times 10 \\ \hline 0 \\ 23 \\ \hline 230 \end{array}$

3. $\begin{array}{r} 26 \\ \times 11 \\ \hline 26 \\ 26 \\ \hline 286 \end{array}$

4. $\begin{array}{r} 22 \\ \times 13 \\ \hline 66 \\ 22 \\ \hline 286 \end{array}$

5. $\begin{array}{r} 44 \\ \times 11 \\ \hline 44 \\ 44 \\ \hline 484 \end{array}$

6. $\begin{array}{r} 23 \\ \times 12 \\ \hline 46 \\ 23 \\ \hline 276 \end{array}$

7. $\begin{array}{r} 12 \\ \times 13 \\ \hline 36 \\ 12 \\ \hline 156 \end{array}$

8. $\begin{array}{r} 20 \\ \times 44 \\ \hline 80 \\ 80 \\ \hline 880 \end{array}$

9. 7,000

10. 1,000

11. 5,000

12. $900 \times 8 = 7{,}200$

13. $600 \times 9 = 5{,}400$

14. $800 \times 7 = 5{,}600$

15.
$$\begin{array}{r} {}^{1\,1} \\ 298 \\ +163 \\ \hline 461 \end{array}$$

16.
$$\begin{array}{r} {}^{1} \\ 562 \\ +475 \\ \hline 1{,}037 \end{array}$$

17.
$$\begin{array}{r} {}^{1} \\ 809 \\ +917 \\ \hline 1{,}726 \end{array}$$

18. $236 + 349 = 585$

$585 - 490 = 95$ plants

19. $12 \times 24 = 288$ pencils

20. $230 \times 3 = 690$ people

Lesson Practice 24A

1. done

2.
$$\begin{array}{r} 24 \\ \times 18 \\ \hline {}^{1\,3} \\ 162 \\ 24 \\ \hline 432 \end{array} \qquad \begin{array}{r} 20+4 \\ \times 10+8 \\ \hline {}^{100}\ \ {}^{30} \\ 100+60+2 \\ 200+40 \\ \hline 400+30+2 \end{array}$$

3.
$$\begin{array}{r} 22 \\ \times 26 \\ \hline {}^{1\,1} \\ 22 \\ 44 \\ \hline 572 \end{array} \qquad \begin{array}{r} 20+2 \\ \times 20+6 \\ \hline {}^{100}\ \ {}^{10} \\ 20+2 \\ 400+40 \\ \hline 500+70+2 \end{array}$$

4.
$$\begin{array}{r} 46 \\ \times 12 \\ \hline {}^{1\,1} \\ 82 \\ 46 \\ \hline 552 \end{array} \qquad \begin{array}{r} 40+6 \\ \times 10+2 \\ \hline {}^{100}\ \ {}^{10} \\ 80+2 \\ 400+60 \\ \hline 500+50+2 \end{array}$$

5.
$$\begin{array}{r} 27 \\ \times 16 \\ \hline {}^{1}{}_{4} \\ {}^{1}22 \\ 27 \\ \hline 432 \end{array} \qquad \begin{array}{r} 20+7 \\ \times 10+6 \\ \hline {}^{100}\ \ {}^{40} \\ {}^{100}\ \ 20+2 \\ 200+70 \\ \hline 400+30+2 \end{array}$$

6.
$$\begin{array}{r} 36 \\ \times 24 \\ \hline {}^{1}{}_{2} \\ {}^{1}24 \\ 62 \\ \hline 864 \end{array} \qquad \begin{array}{r} 30+6 \\ \times 20+4 \\ \hline {}^{100}\ \ {}^{20} \\ {}^{100}\ \ 20+4 \\ 600+20 \\ \hline 800+60+4 \end{array}$$

7.
$$\begin{array}{r} 35 \\ \times 29 \\ \hline {}^{1\,4} \\ 275 \\ {}^{1} \\ 60 \\ \hline 1{,}015 \end{array} \qquad \begin{array}{r} 30+5 \\ \times 20+9 \\ \hline {}^{100}\ \ {}^{40} \\ 200+70+5 \\ {}^{100} \\ 600+\ 0 \\ \hline 1{,}000+\ 0\ +10+5 \end{array}$$

8.
$$\begin{array}{r} 25 \\ \times 23 \\ \hline {}^{1} \\ 65 \\ {}^{1} \\ 40 \\ \hline 575 \end{array} \qquad \begin{array}{r} 20+5 \\ \times 20+3 \\ \hline {}^{10} \\ 60+5 \\ {}^{100} \\ 400 \\ \hline 500+70+5 \end{array}$$

9.
$$\begin{array}{r} 35 \\ \times 15 \\ \hline {}^{1\,2} \\ 155 \\ 35 \\ \hline 525 \end{array} \qquad \begin{array}{r} 30+5 \\ \times 10+5 \\ \hline {}^{100}\ \ {}^{20} \\ 100+50+5 \\ 300+50 \\ \hline 500+20+5 \end{array}$$

525 cars

10.
$$\begin{array}{r} 25 \\ \times 12 \\ \hline {}^{1\,1} \\ 40 \\ 25 \\ \hline 300 \end{array} \qquad \begin{array}{r} 20+5; \\ \times 10+2 \\ \hline {}^{100}\ \ {}^{10} \\ 40+0 \\ 200+50 \\ \hline 300+\ 0\ +0 \end{array}$$

300 tons

Lesson Practice 24B

1.
$$\begin{array}{r} 19 \\ \times 32 \\ \hline {}^{1\,1} \\ 28 \\ {}^{2} \\ 37 \\ \hline 608 \end{array} \qquad \begin{array}{r} 10+9 \\ \times 30+2 \\ \hline {}^{100}\ \ {}^{10} \\ 20+8 \\ {}^{200} \\ 300+70 \\ \hline 600+\ 0\ +8 \end{array}$$

2.
$$\begin{array}{r} 42 \\ \times 66 \\ \hline {}^{1} \\ 242 \\ {}^{1} \\ 242 \\ \hline 2772 \end{array} \qquad \begin{array}{r} 40+2 \\ \times 60+6 \\ \hline {}^{10} \\ 200+40+2 \\ {}^{100} \\ 2000+400+20 \\ \hline 2000+700+70+2 \end{array}$$

3.

33	30+3
×26	×20+6
1 1	100 10
188	100+80+8
66	600+60
858	800+50+8

4.

37	30+7
×12	×10+2
1 1	100 10
64	60+4
37	300+70
444	400+40+4

5.

13	10+3
×19	×10+9
1 2	100 20
97	90+7
13	100+30
247	200+40+7

6.

16	10+6
×29	×20+9
1 5	100 50
94	90+4
22	200+20
464	400+60+4

7.

34	30+4
×17	×10+7
2	20
218	200+10+8
34	300+40
578	500+70+8

8.

48	40+8
×26	×20+6
4	40
248	200+40+8
1	100
86	800+60
1248	1,000+200+40+8

9.

35	30+5; $455
×13	×10+3
1	10
195	100 90+5
35	300+50
455	400+50+5

10.

45	40+5
×24	×20+4
2	20
160	100+60+0
1	100
80	800+ 0
1080	1,000+ 0 +80+0

1,080 mosquitoes

Lesson Practice 24C

1.

23	20+3
×14	×10+4
1	10
182	100 80+2
23	200+30
322	300+20+2

2.

27	20+7
×16	×10+6
1 4	100 40
122	100+20+2
27	200+70
432	400+30+2

3.

29	20+9
×22	×20+2
1 1	100 10
148	100+40+8
48	400+80
638	600+30+8

4.

35	30+5
×15	×10+5
1 2	100 20
155	100 +50+5
35	300+50
525	500+20+5

5.

28	20+8
×22	×20+2
1 1	100 10
146	100 40+6
46	400+60
616	600+10+6

6.

36	30+6
×24	×20+4
2	20
124	100+20+4
1	100
62	600+20
864	800+60+4

7.
$$\begin{array}{r} 44 \\ \underline{\times 27} \\ {\scriptstyle 12} \\ 288 \\ \underline{88} \\ 1188 \end{array} \qquad \begin{array}{r} 40+4 \\ \underline{\times 20+7} \\ {\scriptstyle 100 \quad 20} \\ 200+80+8 \\ \underline{800+80} \\ 1{,}000+100+80+8 \end{array}$$

8.
$$\begin{array}{r} 56 \\ \underline{\times 16} \\ {\scriptstyle 3} \\ 306 \\ \underline{56} \\ 896 \end{array} \qquad \begin{array}{r} 50+6 \\ \underline{\times 10+6} \\ {\scriptstyle 30} \\ 300+\ 0\ +6 \\ \underline{500+60} \\ 800+90+6 \end{array}$$

9.
$$\begin{array}{r} 75 \\ \underline{\times 13} \\ {\scriptstyle 1} \\ 215 \\ \underline{75} \\ 975 \end{array} \qquad \begin{array}{r} 70+5 \\ \underline{\times 10+3} \\ {\scriptstyle 10} \\ 200+10+5 \\ \underline{700+50} \\ 900+70+5 \end{array}$$
; 975 lb

10.
$$\begin{array}{r} 63 \\ \underline{\times 25} \\ {\scriptstyle 1} \\ 305 \\ \underline{126} \\ 1575 \end{array} \qquad \begin{array}{r} 60+3 \\ \underline{\times 20+5} \\ {\scriptstyle 10} \\ 300+\ 0\ +5 \\ \underline{1{,}000+200+60} \\ 1{,}000+500+70+5 \end{array}$$
\$1,575

Systematic Review 24D

1.
$$\begin{array}{r} 32 \\ \underline{\times 17} \\ {\scriptstyle 1} \\ 214 \\ \underline{32} \\ 544 \end{array}$$

2.
$$\begin{array}{r} 14 \\ \underline{\times 28} \\ {\scriptstyle 3} \\ {}^{1}82 \\ \underline{28} \\ 392 \end{array}$$

3.
$$\begin{array}{r} 36 \\ \underline{\times 22} \\ {\scriptstyle 1} \\ {}^{1}62 \\ \underline{62} \\ 792 \end{array}$$

4.
$$\begin{array}{r} 41 \\ \underline{\times 38} \\ 328 \\ \underline{123} \\ 1558 \end{array}$$

5.
$$\begin{array}{r} 50 \\ \underline{\times 32} \\ 100 \\ \underline{150} \\ 1600 \end{array}$$

6.
$$\begin{array}{r} 31 \\ \underline{\times 33} \\ 93 \\ {\scriptstyle 1} \\ \underline{93} \\ 1023 \end{array}$$

7.
$$\begin{array}{r} 12 \\ \underline{\times 41} \\ 12 \\ \underline{48} \\ 492 \end{array}$$

8.
$$\begin{array}{r} 43 \\ \underline{\times 12} \\ 86 \\ {\scriptstyle 1} \\ \underline{43} \\ 516 \end{array}$$

9. $100 \times 4 = 400$

10. $60 \times 8 = 480$

11. $30 \times 7 = 210$

12. 2, 4, 6, 8, 10, 12, 14, 16, 18, 20

13. $3 \times \underline{6} = 18$

14. $3 \times \underline{8} = 24$

15. $8 \times \underline{4} = 32$

16. $15 \times 15 = 225$ sq ft

17. $13 \times 52 = 676$ hours

18. $240 + 335 + 124 = 699$ yd

19. $61 + 57 = 118$
$118 - 59 = 59$ clues

20. $2 \times 5 = 10$
$10 \times 8 = 80$ plants

Systematic Review 24E

1.
$$\begin{array}{r} 18 \\ \times 40 \\ \hline 0 \\ {\scriptstyle 3} \\ 42 \\ \hline 720 \end{array}$$

2.
$$\begin{array}{r} 39 \\ \times 23 \\ \hline {\scriptstyle 12} \\ 197 \\ 68 \\ \hline 897 \end{array}$$

3.
$$\begin{array}{r} 36 \\ \times 99 \\ \hline {\scriptstyle 15} \\ 274 \\ {\scriptstyle 5} \\ 274 \\ \hline 3564 \end{array}$$

4.
$$\begin{array}{r} 76 \\ \times 89 \\ \hline {\scriptstyle 15} \\ 634 \\ {\scriptstyle 14} \\ 568 \\ \hline 6764 \end{array}$$

5.
$$\begin{array}{r} 24 \\ \times 21 \\ \hline 24 \\ {\scriptstyle 1} \\ 48 \\ \hline 504 \end{array}$$

6.
$$\begin{array}{r} 21 \\ \times 14 \\ \hline 84 \\ 21 \\ \hline 294 \end{array}$$

7.
$$\begin{array}{r} 23 \\ \times 22 \\ \hline 46 \\ {\scriptstyle 1} \\ 46 \\ \hline 506 \end{array}$$

8.
$$\begin{array}{r} 12 \\ \times 13 \\ \hline 36 \\ 12 \\ \hline 156 \end{array}$$

9. $600 \times 3 = 1{,}800$
10. $100 \times 5 = 500$
11. $800 \times 8 = 6{,}400$
12. 3, 6, 9, 12, 15, 18, 21, 24, 27, 30
13. $4 \times \underline{9} = 36$
14. $5 \times \underline{5} = 25$
15. $2 \times \underline{6} = 12$
16. $47 \times 31 = 1{,}457$ sq mi
$1{,}457 - 100 = 1{,}357$ sq mi
17. 3 doz + 2 doz = 5 doz
$5 \times 12 = 60$ eggs
18. $11 \times 25 = 275$ cents
19. 300 yards = 900 feet
20. $(7 \times 8) \times 3 = 56 \times 3 = 168$ slices
$7 \times (8 \times 3) = 7 \times 24 = 168$ slices

Systematic Review 24F

1.
$$\begin{array}{r} 41 \\ \times 62 \\ \hline 82 \\ {\scriptstyle 1} \\ 246 \\ \hline 2542 \end{array}$$

2.
$$\begin{array}{r} 55 \\ \times 25 \\ \hline {\scriptstyle 2} \\ 255 \\ {\scriptstyle 1} \\ 100 \\ \hline 1375 \end{array}$$

3.
$$\begin{array}{r} 53 \\ \times 35 \\ \hline {\scriptstyle 11} \\ 255 \\ 159 \\ \hline 1855 \end{array}$$

4.
```
  28
 ×38
 1 6
 164
 2
 64
1064
```

5.
```
  49
 ×38
 1 7
 322
 2
127
1862
```

6.
```
 12
×42
 24
1
48
504
```

7.
```
 24
×22
 48
1
 48
528
```

8.
```
 55
×11
 55
1
55
605
```

9. $200 \times 4 = 800$

10. $300 \times 9 = 2{,}700$

11. $300 \times 3 = 900$

12. 4, 8, 12, 16, 20, 24, 28, 32, 36, 40

13. $10 \times \underline{6} = 60$

14. $4 \times \underline{7} = 28$

15. $3 \times \underline{5} = 15$

16. $2 \times 12 = 24$ cookies

17. $2 \times 18 = 36$
$5 \times 4 = 20$
$20 + 36 = 56$ wheels

18. $10 \times 12 = 120$ books

19. $23 \times 2 = 46$
$3 \times 3 = 9$
$46 + 9 = 55$ points

20. $82 + 140 = 222$ gal
$222 \times 4 = 888$ qt

Lesson Practice 25A

1. done

2.
```
   (600)        624
×   (80)       ×81
(48,000)        624
               1 3
             1 4 862
             5 0,544
```

3.
```
  (300)       305
×  (20)      ×21
(6,000)       305
             610
            6,405
```

4.
```
  (300)       319
×  (30)      ×33
(9,000)       1 2
              937
            1 2
            937
           10,527
```

5.
```
   (500)       495
×   (70)      ×72
(35,000)        1
              1 1 1
               880
           1 6 3
           2835
           35,640
```

6.
```
   (900)       876
×   (20)      ×19
(18,000)        1
              1 6 5
              7234
              876
            16,644
```

7.
```
   (400)       352
×   (30)      ×25
(12,000)        1
               2 1
             1 550
             1
             604
            8,800
```

8.
```
    (700)      681
 ×   (40)     ×38
 (28,000)      16
              4848
              12
              1843
             25,878
```

9.
```
    (500)     500; 9,000 toothpicks
 ×   (20)     ×18
 (10,000)    4000
             500
             9,000
```

10.
```
    (600)      642; 14,766 miles
 ×   (20)      ×23
 (12,000)      11
              1826
              1284
             14,766
```

11.
```
    (200)     212; 5,300 gallons
 ×   (30)     ×25
  (6,000)     11
             1050
             424
             5,300
```

12.
```
    (400)     352; 5,280 pages
 ×   (20)     ×15
  (8,000)     121
             1550
             352
             5,280
```

Lesson Practice 25B

1.
```
    (100)      125
 ×   (30)      ×25
  (3,000)      12
               505
              11
              240
             3,125
```

2.
```
    (700)      681
 ×   (40)      ×38
 (28,000)      16
              4848
              12
              1843
             25,878
```

3.
```
    (500)      492
 ×   (80)      ×75
 (40,000)       11
              2450
             161
             2834
             36,900
```

4.
```
    (100)      145
 ×   (20)      ×15
  (2,000)      2
               525
             1
             145
             2,175
```

5.
```
    (500)      534
 ×   (40)      ×44
 (20,000)      11
              2026
              11
             2026
             23,496
```

6.
```
    (700)      719
 ×  (100)      ×99
 (70,000)      218
              6391
             1 8
             6391
             71,181
```

7.
```
    (200)      154
 ×   (20)      ×16
  (4,000)      3
               624
             1
             154
             2,464
```

8.
```
    (300)      290
 ×   (40)      ×41
 (12,000)      290
              3
              860
             11,890
```

9.
```
    (600)      582
 ×   (70)      ×73
 (42,000)      12
              1546
             151
             3564
             42,486
```

42,486 new mosquitoes

10.

(200)	235
× (10)	×13
(2,000)	1 1
	1 695
	235
	3,055

$3,055

11.

(400)	365
× (40)	×35
(16,000)	1 3
	1525
	1 1
	985
	12,775

12,775 pennies

12.

(400)	365
× (70)	×70
(28,000)	43 0
	2125
	25,550

25,550 pennies

Lesson Practice 25C

1.

(800)	816
× (80)	×79
(64,000)	1 1 5
	7294
	1 4
	5672
	64,464

2.

(500)	492
× (60)	×55
(30,000)	4 1
	1 2050
	4 1
	2050
	27,060

3.

(400)	373
× (60)	×64
(24,000)	1
	2 1
	1282
	1 4 1
	1828
	23,872

4.

(800)	777
× (30)	×33
(24,000)	2 2
	2111
	2 2
	2111
	25,641

5.

(400)	436
× (40)	×36
(16,000)	1
	1 1 3
	2486
	1
	1298
	15,696

6.

(900)	947
× (10)	×14
(9,000)	1 2
	3668
	947
	13,258

7.

(600)	559
× (60)	×63
(36,000)	1
	1 1 2
	1557
	3354
	35,217

8.

(500)	519
× (50)	×52
(25,000)	1
	1028
	4
	2555
	26,988

9.

(1,000)	963; 94,374 sq ft
× (100)	×98
(100,000)	1 4 2
	7284
	1 5 2
	8147
	94,374

10.

(300)	279; $14,508
× (50)	×52
(15,000)	1
	1 1
	448
	1 4
	1355
	14,508

11.
$$\begin{array}{r} (100) \\ \times \quad (10) \\ \hline (1{,}000) \end{array} \qquad \begin{array}{r} 105 \\ \times 12 \\ \hline 210 \\ 105 \\ \hline 1{,}260 \end{array}$$
105; 1,260 pencils

12.
$$\begin{array}{r} (200) \\ \times \quad (10) \\ \hline (2{,}000) \end{array} \qquad \begin{array}{r} 215 \\ \times 14 \\ \hline {\scriptstyle 1\,1\,2} \\ 840 \\ 215 \\ \hline 3{,}010 \end{array}$$
215; 3,010 mi

Systematic Review 25D

1.
$$\begin{array}{r} 873 \\ \times 30 \\ \hline 0 \\ {\scriptstyle 2} \\ 2419 \\ \hline 26{,}190 \end{array}$$

2.
$$\begin{array}{r} 718 \\ \times 38 \\ \hline {\scriptstyle 1\,1\,6} \\ 5684 \\ {\scriptstyle 2} \\ 2134 \\ \hline 27{,}284 \end{array}$$

3.
$$\begin{array}{r} 314 \\ \times 35 \\ \hline {\scriptstyle 2} \\ 1550 \\ {\scriptstyle 1} \\ 932 \\ \hline 10{,}990 \end{array}$$

4.
$$\begin{array}{r} 85 \\ \times 36 \\ \hline {\scriptstyle 1\,3} \\ 480 \\ {\scriptstyle 1\,1} \\ 245 \\ \hline 3{,}060 \end{array}$$

5.
$$\begin{array}{r} 42 \\ \times 59 \\ \hline {\scriptstyle 1} \\ 368 \\ 210 \\ \hline 2{,}478 \end{array}$$

6.
$$\begin{array}{r} 61 \\ \times 47 \\ \hline 427 \\ 244 \\ \hline 2{,}867 \end{array}$$

7. $11 \times 33 = 363$ sq ft
8. $11 + 33 + 11 + 33 = 88$ ft
9. $12 \times 14 = 168$ sq in
10. $12 + 14 + 12 + 14 = 52$ in
11. $25 \times 25 = 625$ sq in
12. $25 \times 4 = 100$ in
13. $42 > 40$
14. $36 = 36$
15. $15 < 16$
16. 5, 10, 15, 20, 25, 30, 35, 40, 45, 50
17. $21 \times 36 = 756$ pieces
18. $10{,}000 \times 4 = \$40{,}000$
19. $43 + 18 = 61$
    $61 - 55 = 6$ cars
20. $12 \times 3 = 36$ feet

Systematic Review 25E

1.
$$\begin{array}{r} 522 \\ \times 93 \\ \hline {\scriptstyle 1\,1} \\ 1566 \\ {\scriptstyle 1\,1} \\ 4588 \\ \hline 48{,}546 \end{array}$$

2.
$$\begin{array}{r} 832 \\ \times 57 \\ \hline {\scriptstyle 1\,2\,1} \\ 5614 \\ {\scriptstyle 1} \\ 4150 \\ \hline 47{,}424 \end{array}$$

3.
$$\begin{array}{r} 471 \\ \times 84 \\ \hline {\scriptstyle 1} \\ {\scriptstyle 1\,2} \\ 1684 \\ {\scriptstyle 5} \\ 3268 \\ \hline 39{,}564 \end{array}$$

4. $$\begin{array}{r} 347 \\ \times\ \ 8 \\ \hline {\scriptstyle 35}\ \ \\ 2426 \\ \hline 2{,}776 \end{array}$$

5. $$\begin{array}{r} 43 \\ \times 25 \\ \hline 215 \\ 86\ \ \\ \hline 1{,}075 \end{array}$$

6. $$\begin{array}{r} 24 \\ \times 31 \\ \hline 24 \\ {\scriptstyle 1}\ \ \ \\ 62\ \ \\ \hline 744 \end{array}$$

7. $21 \times 40 = 840$ sq ft
8. $21 + 40 + 21 + 40 = 122$ ft
9. $7 \times 13 = 91$ sq in
10. $7 + 13 + 7 + 13 = 40$ in
11. $15 \times 15 = 225$ sq in
12. $15 \times 4 = 60$ in
13. $81 > 80$
14. $48 < 50$
15. $66 > 64$
16. 6, 12, 18, 24, 30, 36, 42, 48, 54, 60
17. $1{,}000 \times 6 = 6{,}000$
    $962 \times 6 = 5{,}772$ fish
18. $7 \times 8 = 56$ servings
19. $21 \times 100 = 2{,}100$ lb
20. $455 \times \$15 = \$6{,}825$
    $455 - 35 = 420$ people

Systematic Review 25F

1. $$\begin{array}{r} 712 \\ \times 65 \\ \hline {\scriptstyle 1\ \ 1}\ \ \\ 3550 \\ {\scriptstyle 1}\ \ \ \ \\ 4262\ \ \\ \hline 46{,}280 \end{array}$$

2. $$\begin{array}{r} 360 \\ \times 58 \\ \hline {\scriptstyle 4}\ \ \ \\ 2480 \\ {\scriptstyle 1\ 3}\ \ \ \ \ \\ 1500\ \ \\ \hline 20{,}880 \end{array}$$

3. $$\begin{array}{r} 252 \\ \times 38 \\ \hline {\scriptstyle 1\ 4\ 1}\ \ \\ 1606 \\ {\scriptstyle 1}\ \ \ \ \ \\ 656\ \ \\ \hline 9{,}576 \end{array}$$

4. $$\begin{array}{r} 432 \\ \times\ \ 7 \\ \hline {\scriptstyle 1\ 2\ 1}\ \ \\ 2814 \\ \hline 3{,}024 \end{array}$$

5. $$\begin{array}{r} 41 \\ \times 15 \\ \hline 205 \\ 41\ \ \\ \hline 615 \end{array}$$

6. $$\begin{array}{r} 62 \\ \times 23 \\ \hline 186 \\ 124\ \ \\ \hline 1{,}426 \end{array}$$

7. $16 \times 38 = 608$ sq ft
8. $16 + 38 + 16 + 38 = 108$ ft
9. $10 \times 15 = 150$ sq in
10. $10 + 15 + 10 + 15 = 50$ in
11. $12 \times 12 = 144$ sq in
12. $12 \times 4 = 48$ in
13. $36 = 36$
14. $144 < 150$
15. $22 = 22$
16. 7, 14, 21, 28, 35, 42, 49, 56, 63, 70
17. $98 \times 12 = 1{,}176$ months
18. $11 \times 4 = 44$
    $50 - 44 = 6$ quarters
19. $400 + 200 = 600$
    $375 + 166 = 541$ copies
20. $(4 \times 7) \times 24 = 28 \times 24 = 672$ hours
    $4 \times (7 \times 24) = 4 \times 168 = 672$ hours

Lesson Practice 26A

1. done
2. 1×12
   2×6
   3×4
3. 1×4
   2×2
4. 1×15
   3×5
5. 1×10
   2×5
6. 1×21
   3×7
7. 1×14
   2×7
8. 1×9
   3×3
9. 1×6
   2×3
10. $25 \times 8 = 200$ pennies
11. $25 \times 12 = 300$ cents
12. $25¢ \times 3 = 75¢$
13. 1×12
    2×6
    3×4
14. 1×16
    2×8
    4×4
15. 1×18
    2×9
    3×6

Lesson Practice 26B

1. 1×16
   2×8
   4×4
2. 1×18
   2×9
   3×6
3. 22×1
   11×2
4. 1×6
   2×3
5. 1×14
   2×7
6. 1×15
   3×5
7. 1×8
   2×4
8. 1×12
   2×6
   3×4
9. 1×10
   2×5
10. $9 \times 25 = 225$ pennies
11. $16 \times 25 = 400¢$
12. $35 \times 25 = 875¢$
13. 1×20
    2×10
    4×5
14. 1×21
    3×7
15. 1×24
    2×12
    3×8
    4×6

Lesson Practice 26C

1. 1×22
   2×11
2. 1×8
   2×4
3. 6×1
   2×3
4. 1×14
   2×7
5. 1×21
   3×7
6. 1×10
   2×5
7. 1×18
   2×9
   3×6
8. 1×9
   3×3

9. 1×4
   2×2
10. $4 \times 25 = 100$ pennies
11. $2 \times 7 = 14$
    $14 \times 25 = 350$ cents
12. $5 \times 25¢ = \$1.25$; yes
13. 1×16
    2×8
    4×4
14. three
    1×20
    2×10
    4×5
15. 1×12
    2×6
    3×4

Systematic Review 26D

1. 1×15
   3×5
2. 1×9
   3×3
3. 1×4
   2×2
4. $11 \times 25 = 275$
5. $6 \times 25 = 150$
6. $21 \times 25 = 525$
7. $$\begin{array}{r} 423 \\ \times 57 \\ \hline {}^{12} \\ 2841 \\ {}^{1\,1} \\ 2005 \\ \hline 24{,}111 \end{array}$$
8. $$\begin{array}{r} 276 \\ \times 12 \\ \hline {}^{1} \\ {}^{1\,1\,1} \\ 442 \\ 276 \\ \hline 3{,}312 \end{array}$$
9. $$\begin{array}{r} 614 \\ \times 32 \\ \hline 1228 \\ 1 \\ 1832 \\ \hline 19{,}648 \end{array}$$
10. $$\begin{array}{r} 134 \\ \times \ \ 4 \\ \hline {}^{1\,1} \\ 426 \\ \hline 536 \end{array}$$
11. $$\begin{array}{r} 74 \\ \times 33 \\ \hline {}^{1} \\ 212 \\ {}^{1} \\ 212 \\ \hline 2442 \end{array}$$
12. $$\begin{array}{r} 51 \\ \times 16 \\ \hline 306 \\ 51 \\ \hline 816 \end{array}$$
13. $$\begin{array}{r} 397 \\ -\ 63 \\ \hline 334 \end{array}$$
14. $$\begin{array}{r} \overset{3}{\not 4}\,{}^{1}11 \\ -\ 350 \\ \hline 61 \end{array}$$
15. $$\begin{array}{r} \overset{5}{\not 6}\,{}^{1}\overset{9}{\not 0}\,{}^{1}0 \\ -\ 104 \\ \hline 496 \end{array}$$
16. $\frac{1}{2} = \frac{2}{4} = \frac{3}{6} = \frac{4}{8} = \frac{5}{10}$
17. 1×20
    2×10
    4×5
18. $35 \times \$22 = \$770$
19. $231 \times 2 = 462$ cars
20. $13 + 6 = 19$
    $19 \times 25 = 475$ cents or pennies

Systematic Review 26E

1. 1×6
   2×3
2. 1×12
   2×6
   3×4
3. 1×21
   3×7
4. $7\times25=175$
5. $10\times25=250$
6. $15\times25=375$
7. $$\begin{array}{r} 125 \\ \times54 \\ \hline {}^{2} \\ {}^{1}480 \\ 525 \\ \hline 6{,}750 \end{array}$$
8. $$\begin{array}{r} 731 \\ \times18 \\ \hline 12 \\ 5648 \\ 731 \\ \hline 13{,}158 \end{array}$$
9. $$\begin{array}{r} 378 \\ \times49 \\ \hline 1 \\ 267 \\ 2732 \\ 23 \\ 1282 \\ \hline 18{,}522 \end{array}$$
10. $$\begin{array}{r} 276 \\ \times\ \ 2 \\ \hline 11 \\ 442 \\ \hline 552 \end{array}$$
11. $$\begin{array}{r} 38 \\ \times12 \\ \hline 1 \\ {}^{1}66 \\ 38 \\ \hline 456 \end{array}$$
12. $$\begin{array}{r} 25 \\ \times39 \\ \hline 14 \\ 185 \\ 1 \\ 65 \\ \hline 975 \end{array}$$
13. $$\begin{array}{r} 4 \\ 5\not{5}\,{}^{1}4 \\ -\ \ 27 \\ \hline 527 \end{array}$$
14. $$\begin{array}{r} 976 \\ -763 \\ \hline 213 \end{array}$$
15. $$\begin{array}{r} 3\ {}^{1}7\ 1 \\ \not{4}\,\not{8}\,3 \\ -299 \\ \hline 184 \end{array}$$
16. $\frac{2}{3}=\frac{4}{6}=\frac{6}{9}=\frac{8}{12}=\frac{10}{15}$
17. $321\times3=963$ people
18. $300\times3=900$ feet
19. two
    1×15
    3×5
20. $11+3=14$
    $14-5=9$ times

Systematic Review 26F

1. 1×24
   2×12
   3×8
   4×6
2. 1×16
   2×8
   4×4
3. 1×10
   2×5
4. $2\times25=50$
5. $17\times25=425$
6. $20\times25=500$

7.
$$\begin{array}{r} 249 \\ \times 12 \\ \hline {\scriptstyle 1\,1} \\ 488 \\ 249 \\ \hline 2{,}988 \end{array}$$

8.
$$\begin{array}{r} 218 \\ \times 75 \\ \hline {\scriptstyle 1\,1\,4} \\ 1050 \\ {\scriptstyle 5} \\ 1476 \\ \hline 16{,}350 \end{array}$$

9. $862 \times 10 = 8620$

10.
$$\begin{array}{r} 172 \\ \times\ \ 6 \\ \hline {\scriptstyle 4\,1} \\ 622 \\ \hline 1{,}032 \end{array}$$

11.
$$\begin{array}{r} 92 \\ \times 17 \\ \hline {\scriptstyle 1} \\ 634 \\ 92 \\ \hline 1{,}564 \end{array}$$

12.
$$\begin{array}{r} 56 \\ \times 24 \\ \hline 224 \\ 112 \\ \hline 1{,}344 \end{array}$$

13.
$$\begin{array}{r} 276 \\ -\ \ 12 \\ \hline 264 \end{array}$$

14.
$$\begin{array}{r} {\scriptstyle 4\ {}^{1}4} \\ \not{5}\,\not{5}\,{}^{1}4 \\ -\,396 \\ \hline 158 \end{array}$$

15.
$$\begin{array}{r} {\scriptstyle 6} \\ 6\,\not{7}\,{}^{1}2 \\ -\,325 \\ \hline 347 \end{array}$$

16. $\frac{5}{6} = \frac{10}{12} = \frac{15}{18} = \frac{20}{24} = \frac{25}{30}$

17. $24 \times 4 = 96$ jars

18. $12 \times 4 = 48$ quarters

19. $125 \times 45 = 5{,}625$ sq ft

20. $125 + 45 + 125 = 295$ ft

Lesson Practice 27A

1. done
2. two hundred sixty-one million, eight hundred twenty-nine thousand, one hundred thirty
3. done
4. 42,316
5. 149,273
6. 2,134,911
7. done
8. $100{,}000{,}000 + 50{,}000{,}000 + 900{,}000 + 40{,}000 + 1{,}000 + 200 + 20$
9. $600{,}000{,}000 + 400{,}000 + 90$
10. $6 \times 16 = 96$
11. $10 \times 16 = 160$
12. $13 \times 16 = 208$
13. $100 \times 16 = 1{,}600$ ounces
14. $8 \times 16 = 128$ ounces
15. $2 \times 16 = 32$
    $32 > 30$
    The two-pound can is heavier.

Lesson Practice 27B

1. sixteen million, seven hundred four thousand, nine hundred
2. three hundred twenty-one million, nine hundred fifty-four thousand
3. 4,380
4. 349,622
5. 2,461,800
6. 900,001,373
7. $10{,}000{,}000 + 1{,}000{,}000 + 600{,}000 + 90{,}000 + 1{,}000$
8. $500{,}000{,}000 + 9{,}000{,}000 + 400{,}000 + 30{,}000 + 2{,}000 + 5$
9. $400{,}000{,}000 + 50{,}000{,}000 + 1{,}000{,}000 + 600{,}000 + 90{,}000 + 8{,}000 + 100 + 20 + 3$
10. $9 \times 16 = 144$
11. $11 \times 16 = 176$
12. $34 \times 16 = 544$
13. $5 \times 16 = 80$ ounces

14. $7 \times 16 = 112$ ounces
15. $23 \times 16 = 368$ ounces

Lesson Practice 27C

1. three hundred eighteen million, six hundred eleven thousand, three hundred fifty-three
2. one hundred twenty-six thousand, nine hundred thirty-two
3. 23,914
4. 75,154,900
5. 6,000,342
6. 915,412,965
7. $300{,}000{,}000 + 20{,}000{,}000 + 1{,}000{,}000 + 600{,}000 + 10{,}000 + 8{,}000$
8. $30{,}000{,}000 + 500{,}000 + 800$
9. $100{,}000{,}000 + 1{,}000{,}000 + 7{,}000 + 3$
10. $4 \times 16 = 64$
11. $12 \times 16 = 192$
12. $51 \times 16 = 816$
13. $10 \times 16 = 160$ ounces
14. $14 \times 16 = 224$ ounces
15. $212 \times 16 = 3{,}392$ ounces

Systematic Review 27D

1. ten million, six hundred fifty thousand, three hundred
2. 632,178,431
3. $400{,}000{,}000 + 50{,}000{,}000 + 6{,}000{,}000 + 700{,}000 + 80{,}000 + 9{,}000$
4. 1×14
2×7
5. 1×18
2×9
3×6
6. 1×24
2×12
3×8
4×6
7. $15 \times 16 = 240$
8. $19 \times 25 = 475$
9. $9 \times 3 = 27$
10. $$\begin{array}{r} 123 \\ \underline{\times 67} \\ {\scriptstyle 1} \\ {\scriptstyle 12} \\ {}^{1}741 \\ {\scriptstyle 11} \\ \underline{628} \\ 8{,}241 \end{array}$$
11. $$\begin{array}{r} 147 \\ \underline{\times 51} \\ 147 \\ {\scriptstyle 2} \\ \underline{535} \\ 7{,}497 \end{array}$$
12. $$\begin{array}{r} 38 \\ \underline{\times 15} \\ {\scriptstyle 14} \\ 150 \\ \underline{38} \\ 570 \end{array}$$
13. $\frac{2}{8} = \frac{4}{16} = \frac{6}{24} = \frac{8}{32} = \frac{10}{40} = \frac{12}{48} = \frac{14}{56} = \frac{16}{64} = \frac{18}{72} = \frac{20}{80}$
14. $7 \times 9 = 63$ starters
15. $16 \times 4 = 64$ quarters
16. $6 + 8 = 14$
$15 - 14 = 1$ egg
17. $30 \times 10 = 300$ miles estimated
$29 \times 12 = 348$ miles exact
18. $5 \times 16 = 80$
$80 + 6 = 86$ ounces

Systematic Review 27E

1. $300{,}000{,}000 + 50{,}000{,}000 + 6{,}000{,}000$
2. 784,900,000
3. four hundred million, ninety-eight
4. 1×16
2×8
4×4
5. 1×10
2×5
6. 1×6
2×3
7. $3 \times 16 = 48$
8. $20 \times 4 = 80$

9. $13 \times 3 = 39$

10.
$$\begin{array}{r} 557 \\ \times\ 3 \\ \hline {\scriptstyle 1\,2} \\ 1551 \\ \hline 1{,}671 \end{array}$$

11.
$$\begin{array}{r} 137 \\ \times 59 \\ \hline {\scriptstyle 1} \\ {\scriptstyle 2\,6} \\ 973 \\ {\scriptstyle 2} \\ {\scriptstyle 1\,3} \\ 555 \\ \hline 8{,}083 \end{array}$$

12.
$$\begin{array}{r} 873 \\ \times 21 \\ \hline {\scriptstyle 1} \\ {\scriptstyle 1}873 \\ {\scriptstyle 1} \\ 1646 \\ \hline 18{,}333 \end{array}$$

13. $\frac{3}{9} = \frac{6}{18} = \frac{9}{27} = \frac{12}{36} = \frac{15}{45} =$
$\frac{18}{54} = \frac{21}{63} = \frac{24}{72} = \frac{27}{81} = \frac{30}{90}$

14. $36 \times 43 = 1{,}548$ cans

15. $20 - 12 = 8$
$8 \times 16 = 128$ ounces

16. $\$200 \times 4 = \$800$
$\$175 \times 4 = \$700$

17. $\$175 \times 52 = \$9{,}100$

18. $95 + 68 + 84 + 73 + 91 = 411$ books

Systematic Review 27F

1. $400{,}000{,}000 + 400{,}000 + 400$
2. 132,672,547
3. six hundred ninety-eight million
4. 1×9
   3×3
5. 1×15
   3×5
6. 1×22
   2×11
7. $20 \times 16 = 320$
8. $7 \times 8 = 56$
9. $7 \times 4 = 28$

10.
$$\begin{array}{r} 412 \\ \times 24 \\ \hline 1648 \\ 824 \\ \hline 9{,}888 \end{array}$$

11.
$$\begin{array}{r} 628 \\ \times 41 \\ \hline 628 \\ {\scriptstyle 1\,3} \\ 2482 \\ \hline 25{,}748 \end{array}$$

12.
$$\begin{array}{r} 17 \\ \times 52 \\ \hline {\scriptstyle 1} \\ 24 \\ {\scriptstyle 3} \\ 55 \\ \hline 884 \end{array}$$

13. $\frac{4}{10} = \frac{8}{20} = \frac{12}{30} = \frac{16}{40} = \frac{20}{50} =$
$\frac{24}{60} = \frac{28}{70} = \frac{32}{80} = \frac{36}{90} = \frac{40}{100}$

14. $11 \times 13 = 143$ rolls

15. $A = 12 \times 12 = 144$ sq ft
$P = 12 + 12 + 12 + 12 = 48$ ft

16. $30 - 24 = 6$ miles

17. $(5 \times 24) \times 12 = 120 \times 12 = 1{,}440$ ounces
$5 \times (24 \times 12) = 5 \times 288 = 1{,}440$ ounces

18. $95 + 42 + 68 + 81 = 286$ points

Lesson Practice 28A

1. done

2.
$$\begin{array}{r} 284 \\ \times 362 \\ \hline {\scriptstyle 1} \\ {\scriptstyle 1} \\ {\scriptstyle 1\,1}468 \\ {\scriptstyle 4\,2} \\ 1284 \\ {\scriptstyle 2\,1} \\ 642 \\ \hline 102{,}808 \end{array} \qquad \begin{array}{r} 300 \\ \times 400 \\ \hline 120{,}000 \end{array}$$

3.
```
   880        900
 × 153      × 200
     2    180,000
 1 2440
  4
 4000
 880
134,640
```

4.
```
   714        700
 × 602      × 600
  1428    420,000
  2
 4264
429,828
```

5.
```
 1602     2000
 ×  5     ×  5
 3       10,000
 5010
8,010
```

6.
```
  1768     2000
 ×  12     ×  10
   1 1    20,000
 1 1 1 1
   2426
 1768
21,216
```

7.
```
     8172         8000
    × 354        × 400
     1 1 2   3,200,000
    32488
     3 1
   40550
     2
 24316
2,892,888
```

8.
```
       4 675         5000
       × 292        × 300
         1 1    1,500,000
   1 2   1 1 1
       8 240
   5 6 4
  3 6 4 35
  1 1 1
  8 2 4 0
1, 3 6 5,100
```

9. $120 \times 325 = 39{,}000$ pages

10. $1{,}440 \times 7 = 10{,}080$ minutes

Lesson Practice 28B

1.
```
    325       300
  × 213     × 200
     1 1   60,000
    965
   1
   325
   1
 640
69,225
```

2.
```
    162       200
  × 548     × 500
     1    100,000
   1 4 1
    886
  2
  448
 3 1
 500
88,776
```

3.
```
    536       500
  × 134     × 100
     1     50,000
   1 1 2
   2024
    1
 1 1598
 536
71,824
```

4.
```
    322        300
  × 725      × 700
  1 1 1 1  210,000
    1500
    644
  1 1
 2144
233,450
```

5.
```
 6424      6000
 ×  4      ×  4
 1  1     24,000
 24686
25,696
```

6.
```
   3445        3000
  ×  93        ×  90
     1      270,000
 1 2 1 1 1
    9225
  3 3 4
 27665
320,385
```

7.
```
     1
  12313
   5627        6000
  ×315        ×300
  25005    1,800,000
  5627
  1 2
 15861
1,772,505
```

8.
```
    1
    111
   3579        4000
  ×462        ×500
   6048    2,000,000
  ²345
 ¹18024
  223
 12086
1,653,498
```

9. $\$2,450 \times 12 = \$29,400$

10. $1,350 \times 396 = 534,600$ times

Lesson Practice 28C

1.
```
   1
  113
  627        600
 ×450       ×500
 3005     300,000
   2
 2488
282,150
```

2.
```
  334        300
 ×702       ×700
 ¹668     210,000
  22
 2118
234,468
```

3.
```
   1
  234        200
 ×121       ×100
 ¹234      20,000
 468
234
28,314
```

4.
```
 1114
  415        400
 ×378       ×400
  3280    160,000
    3
  2875
  1
 1235
156,870
```

5.
```
   1
 3567       4000
 ×  8       ×  8
  445      32,000
 24086
 28,536
```

6.
```
   1
  11 2
 2317       2000
 × 64       × 60
  8248    120,000
 1  4
 12862
148,288
```

7.
```
  11
 6536       7000
 ×121       ×100
  6536    700,000
 1  1
 12062
 6536
790,856
```

8.
```
  111
 1562       2000
 ×231       ×200
  1562    400,000
  11
 3586
 11
2024
360,822
```

9. $3,600 \times 24 = 86,400$ seconds

10. $1,260 \times 845 = 1,064,700$ sq ft

Systematic Review 28D

1.
$$\begin{array}{r} 125 \\ \times 306 \\ \hline {}^{1\,1\,3} \\ 620 \\ {}^{1}\quad \\ 365\quad \\ \hline 38{,}250 \end{array} \qquad \begin{array}{r} 100 \\ \times 300 \\ \hline 30{,}000 \end{array}$$

2.
$$\begin{array}{r} 7256 \\ \times \;\; 43 \\ \hline {}^{1\,1\,1\,1} \\ {}^{1\,1} \\ 21658 \\ {}^{2\,2}\quad \\ 28804\quad \\ \hline 312{,}008 \end{array} \qquad \begin{array}{r} 7000 \\ \times \;\; 40 \\ \hline 280{,}000 \end{array}$$

3.
$$\begin{array}{r} 8761 \\ \times 280 \\ \hline 0 \\ {}^{1\,1\,1\,1}\quad \\ {}^{5\,4}\quad \\ 64688\quad \\ {}^{1\,1}\qquad \\ 16422\qquad \\ \hline 2{,}453{,}080 \end{array} \qquad \begin{array}{r} 9000 \\ \times 300 \\ \hline 2{,}700{,}000 \end{array}$$

4. $100{,}000{,}000 + 20{,}000{,}000 +$
$3{,}000{,}000 + 600{,}000$
5. 8×1
4×2
6. 1×4
2×2
7. 1×12
2×6
3×4
8. 1×24
2×12
3×8
4×6
9. $8 \times \underline{8} = 64$
10. $7 \times \underline{6} = 42$
11. $3 \times \underline{1} = 3$
12. $9 \times \underline{5} = 45$
13. $100 + 20 = 120$ yd
$120 \times 3 = 360$ ft
14. $3 \times 2 = 6$ lb
15. $5 \times 3 = 15$ tsp
16. $325 \times 60 = 19{,}500$ gallons
17. $3{,}572 \times \$213 = \$760{,}836$
18. $144 + 50 + 203 = 397$ pipers

Systematic Review 28E

1.
$$\begin{array}{r} 433 \\ \times 127 \\ \hline {}^{1\,1\,2\,2} \\ 2811 \\ 866\quad \\ 433\qquad \\ \hline 54{,}991 \end{array} \qquad \begin{array}{r} 400 \\ \times 100 \\ \hline 40{,}000 \end{array}$$

2.
$$\begin{array}{r} 8192 \\ \times \;\; 74 \\ \hline {}^{1}\quad \\ {}^{1\,1\,1\,3} \\ 32468 \\ {}^{6\,1}\quad \\ 56734\quad \\ \hline 606{,}208 \end{array} \qquad \begin{array}{r} 8000 \\ \times \;\; 70 \\ \hline 560{,}000 \end{array}$$

3.
$$\begin{array}{r} 6123 \\ \times 245 \\ \hline {}^{1\,2\,1\,1} \\ 30505 \\ {}^{1\;\;1}\quad \\ 24482\quad \\ 12246\qquad \\ \hline 1{,}500{,}135 \end{array} \qquad \begin{array}{r} 6000 \\ \times 200 \\ \hline 1{,}200{,}000 \end{array}$$

4. nine million, five hundred fifty-one thousand
5. 1×22
2×11
6. 1×18
2×9
3×6
7. 1×14
2×7
8. 1×10
2×5
9. $9 \times \underline{9} = 81$
10. $6 \times \underline{9} = 54$
11. $4 \times \underline{5} = 20$
12. $8 \times \underline{0} = 0$
13. $125 \times 1{,}200 = 150{,}000$ pounds
14. $125 \times 16 = 2{,}000$ ounces
15. $43 \times 12 = 516$ stamps
16. $12 \times 15 = 180$ tiles

17. $180 \times \$2 = \$360$

18. $18 + 26 = 44$
$44 - 29 = 15$ stories

Systematic Review 28F

1.
$$\begin{array}{r} 156 \\ \times 523 \\ \hline {\scriptstyle 1\ \ 11} \\ 358 \\ {\scriptstyle 1\ 1\ \ } \\ 202\ \ \\ {\scriptstyle 2\ 3\ \ \ \ } \\ 550\ \ \ \ \\ \hline 81{,}588 \end{array} \qquad \begin{array}{r} 200 \\ \times 500 \\ \hline 100{,}000 \end{array}$$

2.
$$\begin{array}{r} 7481 \\ \times\ \ 27 \\ \hline {\scriptstyle 2\ \overset{1}{2}\ 5\ \ } \\ 49867 \\ {\scriptstyle 1\ \ 1\ \ \ \ \ } \\ 14862\ \ \\ \hline 201{,}987 \end{array} \qquad \begin{array}{r} 7000 \\ \times\ \ 30 \\ \hline 210{,}000 \end{array}$$

3.
$$\begin{array}{r} 1222 \\ \times 443 \\ \hline {\scriptstyle 1\ 2\ 2\ 1\ \ } \\ 3666 \\ 4888\ \ \\ 4888\ \ \ \ \\ \hline 541{,}346 \end{array} \qquad \begin{array}{r} 1000 \\ \times 400 \\ \hline 400{,}000 \end{array}$$

4. 65,910,000

5. 1×9
3×3

6. 1×20
2×10
4×5

7. 1×15
3×5

8. 1×21
3×7

9. $7 \times \underline{8} = 56$

10. $8 \times \underline{9} = 72$

11. $5 \times \underline{10} = 50$

12. $6 \times \underline{6} = 36$

13. $13 \times 13 = 169$ apples

14. $60 \times 60 = 3{,}600$ seconds

15. $30 \times 3 = 90$ bows

16. $128 \times 275 = 35{,}200$ ounces

17. $5 + 2 = 7$ pounds
$7 - 3 = 4$ pounds
$4 \times 16 = 64$ ounces

18. $100 \times 20 = 2{,}000$ windows

Lesson Practice 29A

1. done

2. done

3. 1×9
3×3; composite

4. 1×24
2×12
3×8
4×6; composite

5. 1×7; prime

6. 1×15
3×5; composite

7. 1×19; prime

8. 1×6
2×3; composite

9. $5 \times 12 = 60$

10. $10 \times 12 = 120$

11. $2 \times 12 = 24$

12. $9 \times 12 = 108$

13. $4 \times 12 = 48$

14. $7 \times 12 = 84$

15. $12 \times 12 = 144$

16. $1 \times 12 = 12$

17. $8 \times 12 = 96$

18. $3 \times 12 = 36$ eggs

19. $6 \times 12 = 72$ inches

20. $11 \times 12 = 132$ months
Challenge :
12, 24, 36, 48, 60, 72, 84, 96, 108, 120, 132, 144

Lesson Practice 29B

1. 1×2; prime

2. 1×10
2×5; composite

3. 1×17; prime
4. 1×22
   2×11; composite
5. 1×8
   2×4; composite
6. 1×3; prime
7. 1×12
   2×6
   3×4; composite
8. 1×21
   3×7; composite
9. $12\times12=144$
10. $8\times12=96$
11. $6\times12=72$
12. $3\times12=36$
13. $11\times12=132$
14. $4\times12=48$
15. $9\times12=108$
16. $2\times12=24$
17. $10\times12=120$
18. $1\times12=12$ in
19. $5\times12=60$ months
20. $7\times12=84$ people
    Challenge:
    12, 24, 36, 48, 60, 72, 84, 96, 108, 120, 132, 144

Lesson Practice 29C

1. 1×14
   2×7; composite
2. 1×18
   2×9
   3×6; composite
3. 1×5; prime
4. 1×4
   2×2; composite
5. 1×11; prime
6. 1×20
   2×10
   4×5; composite
7. 1×12
   2×6
   3×4; composite
8. 1×23; prime
9. $2\times12=24$
10. $7\times12=84$
11. $11\times12=132$
12. $1\times12=12$
13. $6\times12=72$
14. $5\times12=60$
15. $3\times12=36$
16. $9\times12=108$
17. $12\times12=144$
18. $4\times12=48$ months
19. $10\times12=120$ pencils
20. $8\times12=96$ inches
    Challenge: 12, 24, 36, 48, 60, 72, 84, 96, 108, 120, 132, 144

Systematic Review 29D

1. 1×6
   2×3; composite
2. 1×15
   3×5; composite
3. 1×7; prime
4. $6\times12=72$
5. $3\times12=36$
6. $8\times12=96$
7. $$\begin{array}{r} 45 \\ \times 33 \\ \hline {}^{1} \\ 125 \\ {}^{1} \\ 125 \\ \hline 1{,}485 \end{array}$$
8. $$\begin{array}{r} 4082 \\ \times\ 23 \\ \hline 12246 \\ 8164 \\ \hline 93{,}886 \end{array}$$
9. $$\begin{array}{r} 1499 \\ \times 770 \\ \hline {}^{221}0 \\ {}^{266} \\ 7833 \\ {}^{266} \\ 7833 \\ \hline 1{,}154{,}230 \end{array}$$

10.
```
   8
   9̸¹43
 - 6 50
   2 93
```

11.
```
  ¹3¹2 1
 + 2 79
   6 00
```

12.
```
   476
 +8 13
 1,289
```

13. seventy-six million, eight hundred ninety-three thousand, four hundred twenty
14. $4 \times 21 = 84$ cards
15. $84 \times \$3 = \$252$
16. $163 \times 12 = 1{,}956$ inches
17. $3{,}600 \times 168 = 604{,}800$ seconds
18. $463 + 584 = 1{,}047$ sheep

Systematic Review 29E

1. 1×13; prime
2. 1×16
   2×8
   4×4; composite
3. 1×19; prime
4. $9 \times 12 = 108$
5. $12 \times 12 = 144$
6. $4 \times 12 = 48$
7.
```
       183
     ×644
      3 1
   1 1 422
     3 1
     422
    4 1
    688
   117,852
```
8.
```
      8714
    ×   68
   1 1 1 1
     5  3
    64682
    4  2
   48264
   592,552
```
9.
```
      2408
      ×716
     1 1 1
       2
     12448
     2408
    2
   14856
 1,724,128
```
10.
```
   521
  +765
 1,286
```
11.
```
    0
  21̸¹4
 - 10 8
   10 6
```
12.
```
   1 1
   257
  +463
   720
```
13. $100{,}000{,}000 + 2{,}000{,}000 + 500{,}000 + 700 + 60$
14. $8 \times 10 = 80$
    $8 \times 5 = 40$
    $80 + 40 = 120$ pennies
15. $3{,}500 \times 7 = 24{,}500$ miles
16. $8 + 6 + 8 + 6 = 28$ feet
    $28 \times 12 = 336$ inches
17. $(3 \times 2) \times 7 = 6 \times 7 = \$42$
    $3 \times (2 \times 7) = 3 \times 14 = \$42$
18. $26 + 76 = 102$
    $102 - 8 = 94$ keys

Systematic Review 29F

1. 1×10
   2×5; composite
2. 1×17; prime
3. 1×22
   2×11; composite
4. $11 \times 12 = 132$
5. $10 \times 12 = 120$
6. $5 \times 12 = 60$

7.
$$\begin{array}{r} 964 \\ \times 205 \\ \hline {}^{1}\,3\,2 \\ 4500 \\ 1 \\ 1828 \\ \hline 197{,}620 \end{array}$$

8.
$$\begin{array}{r} 3572 \\ \times\ \ 12 \\ \hline 1\,1\,1 \\ 6044 \\ 3572 \\ \hline 42{,}864 \end{array}$$

9.
$$\begin{array}{r} 6873 \\ \times 343 \\ \hline 2 \\ 1\,1\,1\ 2\,2 \\ 18419 \\ 3\,2\,1 \\ 24282 \\ 2\,2 \\ 18419 \\ \hline 2{,}357{,}439 \end{array}$$

10.
$$\begin{array}{r} 987 \\ -732 \\ \hline 255 \end{array}$$

11.
$$\begin{array}{r} 1\,1 \\ 135 \\ +279 \\ \hline 414 \end{array}$$

12.
$$\begin{array}{r} 1 \\ 862 \\ +345 \\ \hline 1{,}207 \end{array}$$

13. 264,510,000

14. $13+12=25$
$25-7=18$ jars

15. $4\times16=64$ ounces

16. $5\times4=20$
$19+1=20$
enough for her students and herself

17. $(10\times6)\times10=60\times10=600$ tours
$10\times(6\times10)=10\times60=600$ tours

18. $1{,}547\times\$5=\$7{,}735$

Lesson Practice 30A

1. $1\times5{,}280=5{,}280$
2. $1\times2{,}000=2{,}000$
3. done
4. $6\times2{,}000=12{,}000$
5. $8\times5{,}280=42{,}240$
6. $11\times2{,}000=22{,}000$
7. $10\times5{,}280=52{,}800$
8. $4\times2{,}000=8{,}000$
9. $2\times2{,}000=4{,}000$ lb
10. $3\times2{,}000=6{,}000$ lb
    $6{,}000>5{,}000$; no
11. $2\times5{,}280=10{,}560$ ft
12. $12\times5{,}280=$ 63,360 inches

Lesson Practice 30B

1. $1\times5{,}280=5{,}280$
2. $1\times2{,}000=2{,}000$
3. $9\times5{,}280=47{,}520$
4. $10\times2{,}000=20{,}000$
5. $11\times5{,}280=58{,}080$
6. $5\times2{,}000=10{,}000$
7. $16\times5{,}280=84{,}480$
8. $35\times2{,}000=70{,}000$
9. $5\times5{,}280=26{,}400$ ft
10. $22\times2{,}000=44{,}000$ lb
11. $4\times5{,}280=21{,}120$ plants
12. $7\times2{,}000=14{,}000$ lb

Lesson Practice 30C

1. $1\times5{,}280=5{,}280$
2. $1\times2{,}000=2{,}000$
3. $6\times5{,}280=31{,}680$
4. $9\times2{,}000=18{,}000$
5. $13\times5{,}280=68{,}640$
6. $12\times2{,}000=24{,}000$
7. $30\times5{,}280=158{,}400$
8. $123\times2{,}000=246{,}000$
9. $13\times2{,}000=26{,}000$ lb
10. $7\times2{,}000=14{,}000$ lb

11. $6 \times 5{,}280 = 31{,}680$ ft
$31{,}680 > 30{,}000$
Sarah walked farther.

12. $25 \times 5{,}280 = 132{,}000$ ft

Systematic Review 30D

1. $8 \times 5{,}280 = 42{,}240$
2. $4 \times 2{,}000 = 8{,}000$
3. 3×1; prime
4. 1×21
3×7; composite
5. 1×8
2×4; composite
6. 1×11; prime
7. $7 \times 2 = 14$
8. $5 \times 5 = 25$
9. $10 \times 10 = 100$
10. $20 \times 3 = 60$
11.
$$\begin{array}{r} 563 \\ \times 248 \\ \hline {}^{1} \\ {}^{1}\,{}^{42} \\ 4084 \\ {}^{21}\ \\ 2042\ \ \\ {}^{1}\ \ \ \ \\ 1026\ \ \ \\ \hline 139{,}624 \end{array}$$
12.
$$\begin{array}{r} 8657 \\ \times\ \ 15 \\ \hline {}^{1} \\ {}^{323} \\ 40055 \\ 8657\ \\ \hline 129{,}855 \end{array}$$
13.
$$\begin{array}{r} 6214 \\ \times 572 \\ \hline {}^{111} \\ 12428 \\ {}^{1\ 2} \\ 42478\ \\ {}^{1\ 2} \\ 30050\ \ \\ \hline 3{,}554{,}408 \end{array}$$
14. $5 < 8$
15. $56 > 54$
16. $24 = 24$
17. $72 \times 2{,}000 = 144{,}000$ lb
18. $47 \times \$2 = \$94$
19. $11 \times 11 = 121$ fingers
20. $14 - 5 = 9$ feet
$9 + 12 = 21$ feet

Systematic Review 30E

1. $10 \times 5{,}280 = 52{,}800$
2. $8 \times 2{,}000 = 16{,}000$
3. 1×4
2×2; composite
4. 1×18
2×9
3×6; composite
5. 23×1; prime
6. 1×14
2×7; composite
7. $9 \times 3 = 27$
8. $7 \times 4 = 28$
9. $4 \times 4 = 16$
10. $8 \times 8 = 64$
11.
$$\begin{array}{r} 452 \\ \times\ \ 71 \\ \hline {}^{11}452 \\ {}^{31}\ \\ 2854\ \\ \hline 32{,}092 \end{array}$$
12.
$$\begin{array}{r} 1372 \\ \times\ \ 81 \\ \hline {}^{111} \\ 1372 \\ {}^{251}\ \\ 8466\ \\ \hline 111{,}132 \end{array}$$
13.
$$\begin{array}{r} 4912 \\ \times 131 \\ \hline {}^{111} \\ 4912 \\ {}^{2} \\ 12736\ \\ 4912\ \ \\ \hline 643{,}472 \end{array}$$
14. $36 = 36$
15. $49 > 48$

16. $72 < 81$
17. $5{,}280 \times 3 = 15{,}840$ ft
    $15{,}840 < 21{,}000$; yes
18. $12 + 13 = 25$ tons
    $25 \times 2{,}000 = 50{,}000$ pounds
19. $613 \times 466 = 285{,}658$ sq ft
20. $\$68 \times 4 = \$272$
    $\$272 + \$275 = \underline{\$547}$
    $\$700 - \$547 = \underline{\$153}$

Systematic Review 30F

1. $12 \times 5{,}280 = 63{,}360$
2. $11 \times 2{,}000 = 22{,}000$
3. 1×5; prime
4. 20×1
   10×2
   4×5; composite
5. 1×16
   2×8
   4×4; composite
6. 1×19; prime
7. $11 \times 25 = 275$
8. $10 \times 16 = 160$
9. $5 \times 12 = 60$
10. $12 \times 12 = 144$
11.
$$\begin{array}{r} 678 \\ \times\ 125 \\ \hline {\scriptstyle 1\,1\,1} \\ 34 \\ 3050 \\ {\scriptstyle 1\,1} \\ 1246 \\ 678 \\ \hline 84{,}750 \end{array}$$
12.
$$\begin{array}{r} 1563 \\ \times\ 64 \\ \hline {\scriptstyle 1\,1\,1} \\ 221 \\ 4042 \\ {\scriptstyle 3\,3\,1} \\ 6068 \\ \hline 100{,}032 \end{array}$$
13.
$$\begin{array}{r} 6473 \\ \times\ 210 \\ \hline 0 \\ {\scriptstyle 1\;\;1} \\ 6473 \\ {\scriptstyle 1} \\ 12846 \\ \hline 1{,}359{,}330 \end{array}$$
14. $42 > 40$
15. $72 > 70$
16. $27 < 28$
17. $5{,}000 \times 33 = 165{,}000$ ft
18. $5 \times 2{,}000 = 10{,}000$ lb
19. $P = 3 + 3 + 3 + 3 = 12$ ft
    $A = 3 \times 3 = 9$ sq ft
20. $3 \times 12 = 36$ in
    $36 \times 36 = 1{,}296$ sq in

Application and Enrichment Solutions

Application and Enrichment 1G

1. 3×4, 4×3, area = 12
2. 3×7, 7×3, area = 21
3. 6×10, 10×6, area = 60
4. three 4-bars, 3×4, 4×3
5. four 3-bars, 3×4, 4×3
6. two 6-bars, 2×6, 6×2
7. yes

Application and Enrichment 2G

1. $0 \times 7 = 0$
2. $1 \times 5 = 5$ pets
3. 4, 4, yes
4. 4, 3, no
5. 0, 0, yes
6. 3, 4, no
7. 1×4, 4×1, $(1)(4)$, $(4)(1)$, $1 \cdot 4$, $4 \cdot 1$, $1 + 1 + 1 + 1 = 4$

Application and Enrichment 4G

1. Friday
2. Tuesday
3. Monday and Wednesday
4. $2 \times 8 = 16$ quarts

Across

1. product
3. quart
4. rectangle
6. factors

Down

1. pints
2. square
5. area

Application and Enrichment 5G

$5 \times 10 = 50$ flies

Dimes Each Student Has

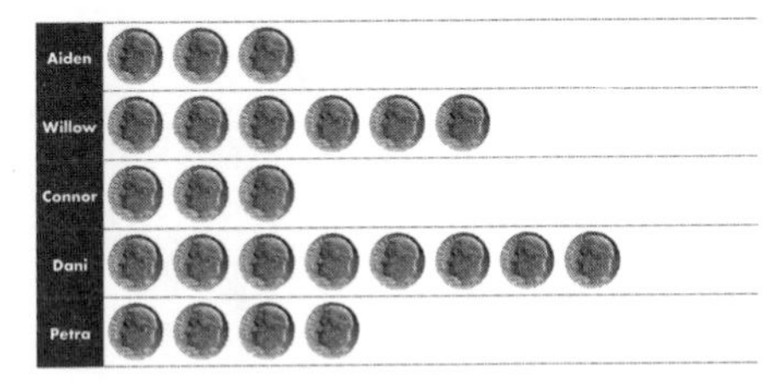

1. Dani
2. Aiden, Conner
3. $10 \times 6 = 60¢$
4. 4 dimes
5. $10 \times 3 = 30¢$
6. Skip counting by ten gives 240¢. You may want to use real dimes and put them in piles to show that this is the same as $2.40.

Application and Enrichment 6G

two fact circle pattern
bottom row of chart:
0, 2, 4, 6, 8, 10, 12, 14, 16, 18, 20

1. 2 times
2. 5 sides
3. pentagon

five fact circle pattern
bottom row of chart:
0, 5, 10, 15, 20, 25, 30, 35, 40, 45, 50
The pattern is a single line.

Application and Enrichment 7G

1. $2 \times 10 = 20$ sq ft
2. $4 \times 5 = 20$ sq ft
3. $1 \times 20 = 20$ sq ft
4. yes

Cost of Each Item in Nickels

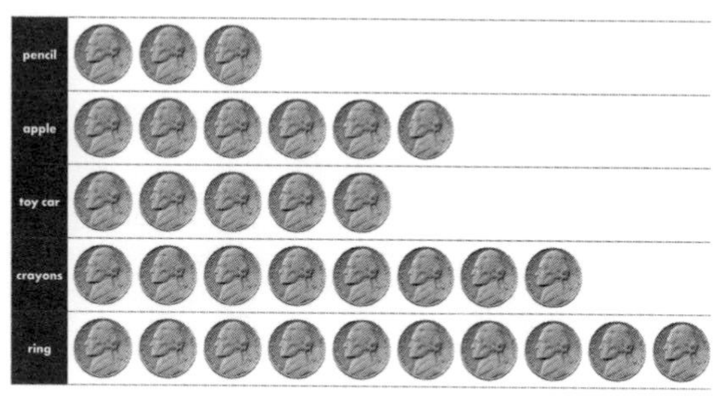

1. ring
2. 40¢
3. pencil
4. 25¢ - 15¢ = 10¢
5. Skip counting by five gives 160¢, You may want to use real nickels and put them in piles to show that this is the same as $1.60.

Application and Enrichment 8G

Students may use any letter they like for the unknown in the problem.

1. 2X = 12, X = 6 boxes
2. 5X = 15, X = 3 windows
3. 2X = 18,
   X = 9 of whatever Rena made
4. $10X = $80, X = 8 items

Logic Problems

1. done
2. Hope - drawing
   Alana - piano
   Jared - juggling

Application and Enrichment 9G

9, 18, 27, <u>36</u> bugs

Application and Enrichment 10G

nine fact circle pattern
bottom row of chart:
0, 9, 18, 27, 36, 45, 54, 63, 72, 81, 90

1. 10 sides
   (A ten-sided shape is a decagon.)
2. counterclockwise

Venn Diagram

1. Region A only - 5, 15, 25, 35, 45
2. Region B only - 60, 70, 80, 90, 100
3. Overlap - 10, 20, 30, 40, 50; 5 numbers
4. 5 numbers

Application and Enrichment 11G

Venn Diagram

1. Region A only: 3, 6, 12, 15, 21, 24, 30
2. Region B only: 36, 45, 54, 63, 72, 81, 90
3. Overlap: 9, 18, 27; 3 numbers
4. 3 numbers

Application and Enrichment 12G

three fact circle pattern
bottom row of chart:
0, 3, 6, 9, 12, 15, 18, 21, 24, 27, 30

1. star
2. 10 points

3, 6, 9, 12, 15, 18, <u>21</u> worms

Application and Enrichment 13G

1. $9 \times 6 = 54$
2. $5 \times 6 = 30$
3. $3 \times 6 = 18$
4. $2 \times 6 = 12$

Application and Enrichment 14G

six fact pattern
bottom row of chart:
0, 6, 12, 18, 24, 30, 36, 42, 48, 54, 60

1. star
2. 5 points

6, 12, 18, 24, 30, 36, 42, 48, 54, 60, 66, 72 nibbles

Application and Enrichment 15G

1. 12 numbers
2. 12 numbers
3. 8 numbers

Logic problems

1. Anna - horse ranch
   Emily - mountains
   James - beach
2. Alex - walked
   Jayna - bus
   Billy - rode in Mom's car

Application and Enrichment 16G

four fact pattern
bottom row of chart:
0, 4, 8, 12, 16, 20, 24, 28, 32, 36

1. star
2. 5 points

4, 8, 12, 16, 20, 24 dollars

Application and Enrichment 17G

$5 \times 5 = 25$
$5 + 5 + 5 + 5 + 5 = 25$
$1 + 3 + 5 + 7 + 9 = 25$

Application and Enrichment 18G

seven fact pattern
bottom row of chart:
0, 7, 14, 21, 28, 35, 42, 49, 56, 63, 70

1. star
2. 10 points

7, 14, 21, 28, 35, 42, 49, 56 days

Application and Enrichment 19G

1. 8, 16, 24, 32, 40, 48
2. 8, 16, 24, 32, 40, 48, 56, 64, 72
3. 8, 16, 24, 32, 40, 48, 56, 64
4. 8, 16, 24, 32, 40, 48, 56

1. Friday
2. Sunday
3. 8, 16, 24, 32
4. 8, 16, 24, 32, 40, 48, 56
5. $0 + 3 + 4 + 6 + 2 + 7 + 5 =$
   27 gallons
6. $0 + 24 + 32 + 48 + 16 + 56 + 40$
   $= 216$ pints
   You may want to separate the column addition in #5 and #6 and then add the parts together.

Application and Enrichment 20G

eight fact circle pattern
bottom row of chart:
0, 8, 16, 24, 32, 40, 48, 56, 64, 72, 80

1. 5
2. pentagon

8, 16, 24, 32, 40, 48, 56, 64 things

Application and Enrichment 21G

1. 20 + 6 = 26
2. 60 + 3 = 63
3. 60 + 4 = 64

$6 \times 6 = 36$
$6 + 6 + 6 + 6 + 6 + 6 = 36$
$1 + 3 + 5 + 7 + 9 + 11 = 36$

$7 \times 7 = 49$
$7 + 7 + 7 + 7 + 7 + 7 + 7 = 49$
$1 + 3 + 5 + 7 + 9 + 11 + 13 = 49$

Application and Enrichment 22G

1. 1 room
2. 4 rooms
3. 8 rooms
4. 12 rooms
5. Write 12 in chart.
6. After Week 1, the numbers of rooms skip count by 4. bottom row of chart: 1, 4, 8, 12, 16, 20, 24

Application and Enrichment 23G

1. $11 \times \$23 = \$253$
2. $11 \times 17 = 187$
3. $11 \times 34 = 374$
4. $11 \times 63 = 693$

1. 2 pt
2. 3 ft
3. 4 qt
4. 5 cents
5. 7 days
6. 8 legs
7. 24 pt
8. 39 ft
9. 44 qt
10. 70 days
11. 88 pt
12. 99 ft

Application and Enrichment 24G

1. 16
2. 24
3. 8 skip count pattern
4. bottom row of chart: 1, 8, 16, 24, 32, 40, 48

Application and Enrichment 25G

Across

2. multiplier
4. ten thousands
5. product
7. hundreds
8. add
9. double

Down

1. thousands
2. multiplicand
3. unit
6. tens

Application and Enrichment 26G

Answers will vary.

1. 4, 8, 22
2. 10, 25, 40
3. 30, 50
4. 33, 51, 69
5. 36, 99, 108, 234

Application and Enrichment 27G

Across

3. dollar
4. ounces
6. two
7. feet

Down

1. gallon
2. tablespoon
3. dime
5. nickel

Area

1. 15 × 20 = 300 sq ft
   12 × 30 = 360 sq ft
   300 + 300 + 360 = 960 sq ft

Application and Enrichment 28G

1. 7 pounds
2. 23 pounds
3. 23 - 7 = 16 pounds
4. first half
5. 16 × 18 = 288 ounces

1. 5 lb
2. Rose
3. 2 × 45 = 90 lb; Seth
4. 130 + 30 = 160 lb
5. 160 < 165; no
6. 130 + 90 = 220 lb; 220 > 165 lb

Application and Enrichment 29G

1. There are three quadrilaterals (4 sides).
2. The square is included as a kind of rectangle.
3. The figure on the bottom right is the square.
4. The figure should have four straight sides and no more than two right angles.

1. quadrilateral, rhombus
2. quadrilateral rectangle
3. quadrilateral
4. quadrilateral, rhombus, square
5. Answers will vary.

Application and Enrichment 30G

A. 3:25
B. 3:40
C. 10:10
D. 10:50

A. 4:13
B. 4:36
C. 9:05
D. 9:21

1. 8:35 + :15 = 8:50
2. 5:55 - :45 = 5:10
3. 3:10 + :35 = 3:45
4. 7:37 - :17 = 7:20

Line plot: Results will vary.

Application and Enrichment A1

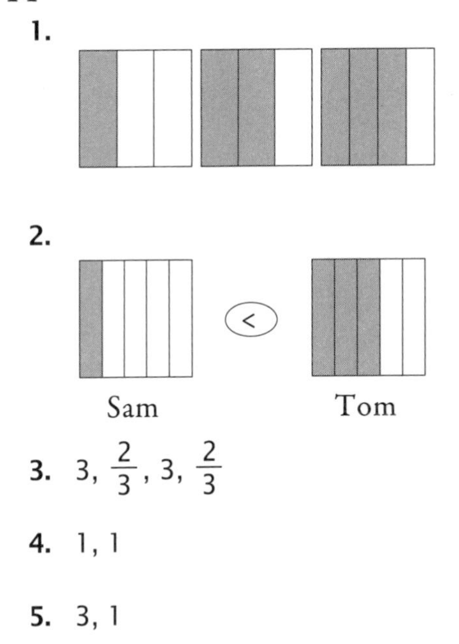

3. 3, $\frac{2}{3}$, 3, $\frac{2}{3}$

4. 1, 1

5. 3, 1

Application and Enrichment B1

Word Problems

1. 5 × 2 = 10 liters
2. 27 + 5 = 32 kilograms

3. about 342 grams
4. There are 4 quarts in a gallon, so she should buy 4 liters of gasoline for each gallon she wants.
5. grams
   kilograms
6. 25 + 25 + 25 + 25 = 100 cm

Test Solutions

Lesson Test 1

1. $4 \times 3 = 12$
   $3 \times 4 = 12$
2. $5 \times 4 = 20$
   $4 \times 5 = 20$
3. $4 \times 2 = 8$
   $2 \times 4 = 8$
4. $2 \times 3 = 6$
   $3 \times 2 = 6$
5. $3 \times 1 = 3$
   $1 \times 3 = 3$
6. $5 \times 3 = 15$
   $3 \times 5 = 15$
7. 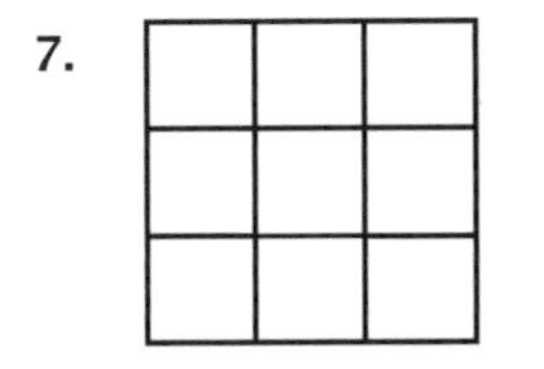
8. 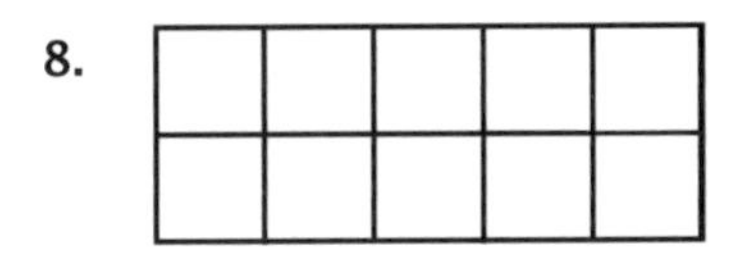

Lesson Test 2

1. $0 \times 2 = 0$
2. $9 \times 1 = 9$
3. $1 \times 7 = 7$
4. $5 \times 1 = 5$
5. $0 \times 9 = 0$
6. $3 \times 0 = 0$
7. $0 \times 7 = 0$
8. $2 \times 1 = 2$
9. $1 \times 6 = 6$
10. $1 \times 1 = 1$
11. $0 \times 1 = 0$
12. $8 \times 1 = 8$
13. $1 \times 3 = 3$
14. $6 \times 0 = 0$
15. $0 \times 5 = 0$
16. $4 \times 0 = 0$
17. $4 \times 1 = 4$
18. $8 \times 0 = 0$
19. $1 \times 9 = 9$
20. $1 \times 5 = 5$
21. $5 = 5$
22. $0 + 0 + 0 + 0 = 0$
23. $7 = 7$
24. $1 \times 6 = 6$ pieces
25. $10 \times 1 = 10$ children

Lesson Test 3

1. 2, 4, 6, 8, 10,
   12, 14, 16, 18, 20
2. 5, 10, 15, 20, 25,
   30, 35, 40, 45, 50
3. 10, 20, 30, 40, 50,
   60, 70, 80, 90, 100
4. $8 \times 1 = 8$
5. $0 \times 5 = 0$
6. $2 \times 1 = 2$
7. $9 \times 0 = 0$
8. $1 \times 7 = 7$
9. $1 \times 0 = 0$
10. $6 \times 1 = 6$
11. $4 \times 0 = 0$
12. $5 + 7 = 12$
13. $14 - 5 = 9$
14. $13 - 6 = 7$
15. $7 + 8 = 15$
16. $0 + 0 + 0 + 0 + 0 = 0$
17. $1 + 1 + 1 + 1 + 1 + 1 + 1 + 1 = 8$
18. 5, 10, $\underline{15}$ books
19. 10, 20, 30, $\underline{40}$ jelly beans
20. 2, 4, 6, 8, 10, 12, 14, $\underline{16}$ apples

Lesson Test 4

1. $9 \times 2 = 18$
2. $2 \times 4 = 8$
3. $1 \times 2 = 2$
4. $2 \times 3 = 6$
5. $6 \times 2 = 12$

6. $2\times5=10$
7. $7\times2=14$
8. $2\times2=4$
9. $8\times2=16$
10. $6\times0=0$
11. $10\times2=20$
12. $3\times1=3$
13. 5, 10, 15, 20, 25, 30, 35, 40, 45, 50
14. $9-4=5$
15. $5+3=8$
16. $17-7=10$
17. $9+8=17$
18. $200+60+3$
19. $2\times6=12$ jars
20. $2\times4=8$ pieces

Lesson Test 5

1. $2\times10=20$
2. $10\times9=90$
3. $3\times10=30$
4. $10\times7=70$
5. $6\times10=60$
6. $10\times1=10$
7. $4\times10=40$
8. $10\times5=50$
9. $10\times8=80$
10. $5\times2=10$
11. $1\times3=3$
12. $8\times2=16$
13. $\begin{array}{r} 34 \\ -21 \\ \hline 13 \end{array}$
14. $\begin{array}{r} 55 \\ +42 \\ \hline 97 \end{array}$
15. $18-1=17$
16. $\begin{array}{r} 60 \\ +17 \\ \hline 77 \end{array}$
17. $100+90+4$
18. $7\times10=70$ cents
19. $10\times2=20$ pt
    $20-10=10$ pt
20. $3\times10=30$ mi

Lesson Test 6

1. $3\times5=15$
2. $5\times7=35$
3. $2\times5=10$
4. $5\times8=40$
5. $6\times5=30$
6. $5\times0=0$
7. $10\times5=50$
8. $5\times5=25$
9. $5\times4=20$
10. $2\times7=14$
11. $9\times2=18$
12. $10\times6=60$
13. $\begin{array}{r} ^{1} \\ 61 \\ +39 \\ \hline 100 \end{array}$
14. $\begin{array}{r} ^{1} \\ 47 \\ +25 \\ \hline 72 \end{array}$
15. $\begin{array}{r} ^{1} \\ 56 \\ +36 \\ \hline 92 \end{array}$
16. $\begin{array}{r} ^{1} \\ 84 \\ +19 \\ \hline 103 \end{array}$
17. $5+5+5+5+5+5+5+5=40$
18. $2\times4=8$
    $5\times3=15$
    $8+15=23$ chores
19. $3\times5=15¢$
    $8\times10=80¢$
    $15+80=95¢$
20. $6\times5=30$ minutes

Unit Test I

1. $2 \times 2 = 4$
2. $7 \times 10 = 70$
3. $10 \times 10 = 100$
4. $5 \times 2 = 10$
5. $0 \times 1 = 0$
6. $10 \times 4 = 40$
7. $2 \times 8 = 16$
8. $1 \times 5 = 5$
9. $5 \times 9 = 45$
10. $4 \times 5 = 20$
11. $1 \times 6 = 6$
12. $3 \times 5 = 15$
13. $1 \times 9 = 9$
14. $10 \times 2 = 20$
15. $3 \times 10 = 30$
16. $5 \times 8 = 40$
17. $0 \times 0 = 0$
18. $10 \times 6 = 60$
19. $6 \times 5 = 30$
20. $2 \times 7 = 14$
21. $3 \times 0 = 0$
22. $2 \times 3 = 6$
23. $7 \times 5 = 35$
24. $2 \times 9 = 18$
25. $6 \times 2 = 12$
26. $10 \times 8 = 80$
27. $4 \times 2 = 8$
28. $5 \times 5 = 25$
29. $10 \times 9 = 90$
30. $0 \times 2 = 0$
31. $7 \times 1 = 7$
32. $10 \times 5 = 50$
33. 2
34. 10
35. 5

Lesson Test 7

1. $5 \times 1 = 5$ sq ft
2. $10 \times 5 = 50$ sq mi
3. $1 \times 2 = 2$ sq ft
4. $10 \times 1 = 10$ sq in
5. $5 \times 2 = 10$ sq in
6. $10 \times 2 = 20$ sq yd
7. $5 \times 6 = 30$
8. $2 \times 8 = 16$
9. $10 \times 7 = 70$
10. $5 \times 5 = 25$
11. $2 \times 4 = 8$ sq ft
12. $3 \times 2 = 6$ sq ft

Lesson Test 8

1. $10 \times \underline{2} = 20$
2. $3 \times \underline{1} = 3$
3. $9 \times \underline{5} = 45$
4. $5 \times \underline{10} = 50$
5. $2 \times \underline{4} = 8$
6. $5 \times \underline{2} = 10$
7. $2 \times \underline{9} = 18$
8. $7 \times \underline{2} = 14$
9. $7 \times 5 = 35$ sq mi
10. $10 \times 10 = 100$ sq in
11. $2 \times 3 = 6$ sq ft
12. $$\begin{array}{r} {}^{1} \\ 39 \\ +52 \\ \hline 91 \end{array}$$
13. $$\begin{array}{r} 8\phantom{{}^{1}5} \\ \cancel{9}\,{}^{1}5 \\ -16 \\ \hline 79 \end{array}$$
14. $$\begin{array}{r} {}^{1} \\ 78 \\ +25 \\ \hline 103 \end{array}$$
15. $$\begin{array}{r} 5\phantom{{}^{1}1} \\ \cancel{6}\,{}^{1}1 \\ -48 \\ \hline 13 \end{array}$$
16. $45 - 37 = 8$ mi
17. $5 \times \underline{8} = 40$; 8 trips
18. $2 \times \underline{6} = 12$; 6 qt
19. $\$5 \times \underline{7} = \$35$; 7 five-dollar bills
20. $10 \times \underline{5} = 50$; 5 steps

Lesson Test 9

1. 9, 18, 27, 36, 45,
   54, 63, 72, 81, 90
2. $\frac{2}{9}=\frac{4}{18}=\frac{6}{27}=\frac{8}{36}=\frac{10}{45}=$
   $\frac{12}{54}=\frac{14}{63}=\frac{16}{72}=\frac{18}{81}=\frac{20}{90}$
3. $\frac{0}{(5)(0)},\frac{5}{(5)(1)},\frac{10}{(5)(2)},\frac{15}{(5)(3)},$
   $\frac{20}{(5)(4)},\frac{25}{(5)(5)},\frac{30}{(5)(6)},\frac{35}{(5)(7)},$
   $\frac{40}{(5)(8)},\frac{45}{(5)(9)},\frac{50}{(5)(10)}$
4. $5\times\underline{9}=45$
5. $8\times\underline{2}=16$
6. $5\times\underline{7}=35$
7. $8\times\underline{10}=80$
8. $8\times1=8$ sq mi
9. $1\times1=1$ sq in
10. $6\times5=30$ sq ft
11. $\begin{array}{r} {}^{1} \\ 32 \\ +58 \\ \hline 90 \end{array}$
12. $\begin{array}{r} {}^{4}\not{5}\,{}^{1}4 \\ -47 \\ \hline 7 \end{array}$
13. $\begin{array}{r} {}^{1} \\ 65 \\ +18 \\ \hline 83 \end{array}$
14. $\begin{array}{r} {}^{6}\not{7}\,{}^{1}0 \\ -21 \\ \hline 49 \end{array}$
15. $4\times10¢=40¢$
    $3\times5¢=15¢$
    $40¢+15¢=55¢$
16. 9, 18, $\underline{27}$ marbles
17. $5\times\$9=\$45$
18. 9, 18, 27, 36, 45, 54, 63, 72, $\underline{81}$ chips

Lesson Test 10

1. $9\times0=0$
2. $9\times9=81$
3. $3\times9=27$
4. $7\times9=63$
5. $9\times4=36$
6. $6\times9=54$
7. $9\times8=72$
8. $2\times9=18$
9. $10\times9=90$
10. $9\times5=45$
11. $2\times6=12$
12. $5\times8=40$
13. $\frac{0}{9\cdot0},\frac{9}{9\cdot1},\frac{18}{9\cdot2},\frac{27}{9\cdot3},\frac{36}{9\cdot4},\frac{45}{9\cdot5},$
    $\frac{54}{9\cdot6},\frac{63}{9\cdot7},\frac{72}{9\cdot8},\frac{81}{9\cdot9},\frac{90}{9\cdot10}$
14. $2\times\underline{9}=18$
15. $5\times\underline{5}=25$
16. $4\times\underline{0}=0$
17. $8\times\underline{9}=72$
18. $9\times3=27$ minutes
19. $9\times9=81$ lives
20. $\$9\times8=\$72$

Lesson Test 11

1. 3, 6, 9, 12, 15, 18, 21, 24, 27, 30
2. $\frac{3}{5}=\frac{6}{10}=\frac{9}{15}=\frac{12}{20}=\frac{15}{25}=$
   $\frac{18}{30}=\frac{21}{35}=\frac{24}{40}=\frac{27}{45}=\frac{30}{50}$
3. $\frac{0}{9\times0},\frac{9}{9\times1},\frac{18}{9\times2},\frac{27}{9\times3},\frac{36}{9\times4},\frac{45}{9\times5},$
   $\frac{54}{9\times6},\frac{63}{9\times7},\frac{72}{9\times8},\frac{81}{9\times9},\frac{90}{9\times10}$
4. $5\times4=20$
5. $2\times7=14$
6. $5\times3=15$
7. $2\times8=16$
8. $\begin{array}{r} {}^{1} \\ 21 \\ 39 \\ +\ 6 \\ \hline 66 \end{array}$

9.
$$\begin{array}{r} 28 \\ 40 \\ +61 \\ \hline 129 \end{array}$$

10.
$$\begin{array}{r} {}^{2} \\ 65 \\ 23 \\ 15 \\ +\ \ 7 \\ \hline 110 \end{array}$$

11.
$$\begin{array}{r} {}^{1} \\ 32 \\ 33 \\ 42 \\ +13 \\ \hline 120 \end{array}$$

12. 3, 6, 9, 12 ribbons

13. 3, 6, 9, 12, 15, 18, 21 meals

14. 3, 6, 9, 12, 15 pages

15. 3, 6, 9, 12, 15, 18 miles

Lesson Test 12

1. $3 \times 9 = 27$
2. $8 \times 3 = 24$
3. $3 \times 3 = 9$
4. $4 \times 3 = 12$
5. $3 \times 2 = 6$
6. $6 \times 3 = 18$
7. $3 \times 10 = 30$
8. $7 \times 3 = 21$
9. $3 \times 1 = 3$
10. $5 \times 3 = 15$
11. $6 \times 9 = 54$
12. $9 \times 9 = 81$

13.
$$\begin{array}{r} {}^{1} \\ 42 \\ 34 \\ +\ \ 8 \\ \hline 84 \end{array}$$

14.
$$\begin{array}{r} {}^{1} \\ 17 \\ 11 \\ +13 \\ \hline 41 \end{array}$$

15.
$$\begin{array}{r} \overset{8}{\cancel{9}}\,{}^{1}2 \\ -\,2\,5 \\ \hline 6\,7 \end{array}$$

16.
$$\begin{array}{r} \overset{7}{\cancel{8}}\,{}^{1}4 \\ -\,3\,6 \\ \hline 4\,8 \end{array}$$

17. $6 \times 3 = 18'$

18. $9 \times 3 = 27$

19. $9 \times 3 = 27'$

20. $3 \times 4 = 12$ sandwiches

Lesson Test 13

1. 6, 12, 18, 24, 30, 36, 42, 48, 54, 60
2. 9, 18, 27, 36, 45, 54, 63, 72, 81, 90
3. $\frac{1}{2} = \frac{2}{4} = \frac{3}{6} = \frac{4}{8} = \frac{5}{10}$
4. $3 \times 3 = 9$
5. $3 \times 9 = 27$
6. $9 \times 9 = 81$
7. $0 \times 0 = 0$
8. $10 + 10 + 13 + 13 = 46$ mi
9. $15 + 32 + 15 + 32 = 94'$
10. $6 + 6 + 6 + 6 = 24''$
11. 6, 12, 18, 24, 30 bulbs
12. 6, 12, 18, 24, 30, 36, 42 letters
13. 6, 12, 18, 24 blocks
14. 6, 12, 18, 24, 30, 36 trips

Lesson Test 14

1. $6 \times 7 = 42$
2. $9 \times 6 = 54$
3. $6 \times 3 = 18$
4. $1 \times 6 = 6$
5. $6 \times 4 = 24$
6. $10 \times 6 = 60$
7. $6 \times 8 = 48$
8. $6 \times 6 = 36$
9. $6 \times 2 = 12$
10. $3 \times 3 = 9$
11. $6 \times 5 = 30$
12. $2 \times 10 = 20$

13. $5 \times 5 = 25$
14. $7 \times 9 = 63$
15. $9 \times 9 = 81$
16. $8 \times 3 = 24$
17. $\frac{5}{6} = \frac{10}{12} = \frac{15}{18} = \frac{20}{24} = \frac{25}{30}$
18. $2 \times 3 = 6$ tsp
19. $6 \times 3 = 18$ ft
20. $6 \times 6 = 36$ cups

Unit Test II

1. $8 \times 6 = 48$
2. $3 \times 5 = 15$
3. $9 \times 6 = 54$
4. $3 \times 7 = 21$
5. $1 \times 9 = 9$
6. $6 \times 0 = 0$
7. $10 \times 9 = 90$
8. $9 \times 8 = 72$
9. $3 \times 2 = 6$
10. $7 \times 9 = 63$
11. $3 \times 1 = 3$
12. $3 \times 6 = 18$
13. $10 \times 3 = 30$
14. $6 \times 3 = 18$
15. $4 \times 9 = 36$
16. $6 \times 4 = 24$
17. $2 \times 6 = 12$
18. $3 \times 3 = 9$
19. $5 \times 6 = 30$
20. $9 \times 2 = 18$
21. $3 \times 9 = 27$
22. $9 \times 9 = 81$
23. $5 \times 9 = 45$
24. $3 \times 8 = 24$
25. $4 \times 3 = 12$
26. $6 \times 6 = 36$
27. $7 \times 6 = 42$
28. $10 \times 6 = 60$
29. $2 \times 9 = 18$ sq in
30. $2 + 9 + 2 + 9 = 22"$
31. $4 \times 3 = 12'$
32. $6 \times 3 = 18$ tsp

Lesson Test 15

1. 4, 8, 12, 16, 20, 24, 28, 32, 36, 40
2. 3, 6, 9, 12, 15, 18, 21, 24, 27, 30
3. $\frac{2}{4} = \frac{4}{8} = \frac{6}{12} = \frac{8}{16} = \frac{10}{20}$
4. $3 \times 4 = 12$
5. $5 \times 6 = 30$
6. $9 \times 9 = 81$
7. $6 \times 7 = 42$
8. $5 \times 2 = 10$
9. $1 \times 8 = 8$
10. $9 \times 7 = 63$
11. $3 \times 3 = 9$
12. 4, 8, 12, 16, 20, $\underline{24}$ qt
13. 4, 8, $\underline{12}$ minutes
14. 4, 8, 12, $\underline{16}$ times
15. 4, 8, 12, 16, 20, 24, $\underline{28}$ hooves

Lesson Test 16

1. $4 \times 9 = 36$
2. $4 \times 4 = 16$
3. $4 \times 2 = 8$
4. $5 \times 4 = 20$
5. $4 \times 3 = 12$
6. $8 \times 4 = 32$
7. $4 \times 7 = 28$
8. $4 \times 6 = 24$
9. $10 \times 4 = 40$
10. $9 \times 6 = 54$
11. $3 \times 7 = 21$
12. $6 \times 6 = 36$
13. $5 \times 5 = 25$ sq in
14. $5 + 5 + 5 + 5 = 20$ in
15. $2 \times \underline{6} = 12$
16. $3 \times \underline{3} = 9$
17. $10 \times \underline{3} = 30$
18. $6 \times 4 = 24$ quarters
19. $5 \times 4 = 20$ plants
20. $3 \times 4 = 12$ gumballs

Lesson Test 17

1. 7, 14, 21, 28, 35, 42, 49, 56, 63, 70
2. $\frac{6}{7}=\frac{12}{14}=\frac{18}{21}=\frac{24}{28}=\frac{30}{35}$
3. $70\times3=210$
4. $40\times2=80$
5. $90\times5=450$
6. $7\times\underline{2}=14$
7. $4\times\underline{9}=36$
8. $6\times\underline{7}=42$
9. $5\times\underline{3}=15$
10. $27<28$
11. $13=13$
12. 4 ft $<$ 30 ft
13. 7, 14, 21, 28, 35, $\underline{42}$ days
14. 7, 14, 21, 28, 35, 42, $\underline{49}$ points
15. 7, 14, $\underline{21}$ pieces

Lesson Test 18

1. $7\times2=14$
2. $4\times7=28$
3. $3\times4=12$
4. $6\times7=42$
5. $7\times9=63$
6. $10\times7=70$
7. $7\times7=49$
8. $5\times7=35$
9. $8\times7=56$
10. $6\times6=36$
11. $50\times5=250$
12. $90\times2=180$
13. $100\times4=400$
14. $400\times2=800$
15. $200\times2=400$
16. $100\times6=600$
17. $4\times7=28$ quarters
18. $\$8\times7=\$56$
19. $7\times6=42'$
20. $7\times100=700$ paper clips

Lesson Test 19

1. 8, 16, 24, 32, 40, 48, 56, 64, 72, 80
2. 7, 14, 21, 28, 35, 42, 49, 56, 63, 70
3. $\frac{1}{5}=\frac{2}{10}=\frac{3}{15}=\frac{4}{20}=\frac{5}{25}$
4. $7\times7=49$
5. $5\times7=35$
6. $6\times5=30$
7. $7\times8=56$
8. $8\times3=24$
9. $0\times9=0$
10. $6\times7=42$
11. $6\times8=48$
12. $\begin{array}{r} 450 \\ +\,106 \\ \hline 556 \end{array}$
13. $\begin{array}{r} {\scriptstyle 1} \\ 392 \\ +\,117 \\ \hline 509 \end{array}$
14. $\begin{array}{r} {\scriptstyle 1} \\ 546 \\ +\,237 \\ \hline 783 \end{array}$
15. 8, 16, 24, 32, 40, $\underline{48}$ pints
16. 8, 16, 24, 32, $\underline{40}$ hours
17. 8, 16, 24, 32, 40, $\underline{48}$ sq ft
18. 8, 16, 24, $\underline{32}$ quarters

Lesson Test 20

1. $8\times3=24$
2. $9\times8=72$
3. $8\times7=56$
4. $4\times8=32$
5. $8\times8=64$
6. $6\times8=48$
7. $8\times10=80$
8. $5\times8=40$
9. $80\times2=160$
10. $7\times6=42$
11. $50\times7=350$
12. $300\times3=900$
13. $9>8$
14. 2 pt $<$ 16 pt

15. $24 = 24$

16. $\begin{array}{r} \scriptstyle 6\ {}^{1}1\ \ \\ \not{7}\,\not{2}\,{}^{1}0 \\ -\,4\,3\,1 \\ \hline 2\,8\,9 \end{array}$

17. $\begin{array}{r} \scriptstyle 7\ {}^{1}8\ \ \\ \not{8}\,\not{9}\,{}^{1}2 \\ -\,6\,9\,5 \\ \hline 1\,9\,7 \end{array}$

18. $\begin{array}{r} \scriptstyle 3\ \ 9\ \ \\ \not{4}\,{}^{1}\not{0}\,{}^{1}6 \\ -\,1\,1\,9 \\ \hline 2\,8\,7 \end{array}$

19. $3 \times 8 = 24$ glasses

20. $8 \times 5 = 40$ words

Unit Test III

1. $4 \times 7 = 28$
2. $5 \times 5 = 25$
3. $0 \times 1 = 0$
4. $8 \times 9 = 72$
5. $3 \times 4 = 12$
6. $3 \times 8 = 24$
7. $5 \times 3 = 15$
8. $7 \times 10 = 70$
9. $3 \times 3 = 9$
10. $10 \times 6 = 60$
11. $8 \times 5 = 40$
12. $6 \times 4 = 24$
13. $8 \times 2 = 16$
14. $4 \times 9 = 36$
15. $7 \times 7 = 49$
16. $4 \times 5 = 20$
17. $6 \times 1 = 6$
18. $7 \times 5 = 35$
19. $3 \times 6 = 18$
20. $10 \times 2 = 20$
21. $9 \times 3 = 27$
22. $10 \times 5 = 50$
23. $7 \times 2 = 14$
24. $7 \times 8 = 56$
25. $10 \times 3 = 30$
26. $4 \times 2 = 8$
27. $8 \times 8 = 64$
28. $10 \times 9 = 90$
29. $5 \times 2 = 10$
30. $4 \times 4 = 16$
31. $8 \times 6 = 48$
32. $5 \times 0 = 0$
33. $10 \times 10 = 100$
34. $6 \times 9 = 54$
35. $10 \times 8 = 80$
36. $9 \times 9 = 81$
37. $6 \times 5 = 30$
38. $8 \times 4 = 32$
39. $9 \times 5 = 45$
40. $6 \times 6 = 36$
41. $2 \times 9 = 18$
42. $10 \times 4 = 40$
43. $6 \times 7 = 42$
44. $2 \times 3 = 6$
45. $6 \times 2 = 12$
46. $9 \times 7 = 63$
47. $2 \times 2 = 4$
48. $7 \times 3 = 21$
49. $5 \times 2 = 10$
50. $8 \times 3 = 24$
51. $6 \times 4 = 24$
52. $10 \times 4 = 40$
53. $4 \times 3 = 12$
54. $7 \times 8 = 56$
55. $6 \times 5 = 30$ sq ft
56. $6 + 5 + 6 + 5 = 22'$

Lesson Test 21

1. $\begin{array}{r} 22 \\ \times\ \ 4 \\ \hline 88 \end{array} \qquad \begin{array}{r} 20+2 \\ \times \qquad 4 \\ \hline 80+8 \end{array}$

2. $\begin{array}{r} 12 \\ \times\ \ 3 \\ \hline 36 \end{array} \qquad \begin{array}{r} 10+2 \\ \times \qquad 3 \\ \hline 30+6 \end{array}$

3. $\begin{array}{r} 11 \\ \times\ \ 5 \\ \hline 55 \end{array} \qquad \begin{array}{r} 10+1 \\ \times \qquad 5 \\ \hline 50+5 \end{array}$

4.
$$\begin{array}{r} 41 \\ \times\ 2 \\ \hline 82 \end{array} \qquad \begin{array}{r} 40+1 \\ \times\quad 2 \\ \hline 80+2 \end{array}$$

5.
$$\begin{array}{r} 211 \\ \times\ 3 \\ \hline 633 \end{array} \qquad \begin{array}{r} 200+10+1 \\ \times\qquad 3 \\ \hline 600+30+3 \end{array}$$

6.
$$\begin{array}{r} 202 \\ \times\ 4 \\ \hline 808 \end{array} \qquad \begin{array}{r} 200+00+2 \\ \times\qquad 4 \\ \hline 800+00+8 \end{array}$$

7.
$$\begin{array}{r} 111 \\ \times\ 7 \\ \hline 777 \end{array} \qquad \begin{array}{r} 100+10+1 \\ \times\qquad 7 \\ \hline 700+70+7 \end{array}$$

8.
$$\begin{array}{r} 112 \\ \times\ 4 \\ \hline 448 \end{array} \qquad \begin{array}{r} 100+10+2 \\ \times\qquad 4 \\ \hline 400+40+8 \end{array}$$

9.
$$\begin{array}{r} {}^{1} \\ 21 \\ 35 \\ +24 \\ \hline 80 \end{array}$$

10.
$$\begin{array}{r} {}^{1} \\ 42 \\ 19 \\ +37 \\ \hline 98 \end{array}$$

11.
$$\begin{array}{r} 2\,\overset{3}{\not{4}}\,{}^{1}5 \\ -\ \ 16 \\ \hline 229 \end{array}$$

12.
$$\begin{array}{r} \overset{2}{\not{3}}\,{}^{1}\overset{9}{\not{0}}\,{}^{1}4 \\ -228 \\ \hline 76 \end{array}$$

13. $103 \times 2 = 206$ ears

14. $12 \times 4 = 48$ eggs

15. $13 \times 3 = 39$ miles

Lesson Test 22

1. 30
2. 40
3. 80
4. 300
5. 500
6. 100
7. 7,000
8. 3,000
9. 8,000
10. $40 \times 3 = 120$
11. $70 \times 2 = 140$
12. $30 \times 4 = 120$
13. $200 \times 7 = 1{,}400$
14. $800 \times 4 = 3{,}200$
15. $200 \times 4 = 800$

16.
$$\begin{array}{r} 24 \\ \times\ 2 \\ \hline 48 \end{array} \qquad \begin{array}{r} 20+4 \\ \times\quad 2 \\ \hline 40+8 \end{array}$$

17.
$$\begin{array}{r} 12 \\ \times\ 3 \\ \hline 36 \end{array} \qquad \begin{array}{r} 10+2 \\ \times\quad 3 \\ \hline 30+6 \end{array}$$

18. \$2,000

19. $\$50 \times 6 = \$300$

20. $200 \times 4 = 800$ miles

Lesson Test 23

1.
$$\begin{array}{r} 23 \\ \times 23 \\ \hline 69 \\ 46 \\ \hline 529 \end{array}$$

2.
$$\begin{array}{r} 22 \\ \times 44 \\ \hline 88 \\ 88 \\ \hline 968 \end{array}$$

3.
$$\begin{array}{r} 21 \\ \times 11 \\ \hline 21 \\ 21 \\ \hline 231 \end{array}$$

4.
$$\begin{array}{r} 32 \\ \times 12 \\ \hline 64 \\ 32 \\ \hline 384 \end{array}$$

5.
$$\begin{array}{r} 17 \\ \underline{\times 11} \\ 17 \\ \underline{17} \\ 187 \end{array}$$

6.
$$\begin{array}{r} 18 \\ \underline{\times 11} \\ 18 \\ \underline{18} \\ 198 \end{array}$$

7.
$$\begin{array}{r} 23 \\ \underline{\times 32} \\ 146 \\ \underline{69} \\ 736 \end{array}$$

8.
$$\begin{array}{r} 20 \\ \underline{\times 44} \\ 80 \\ \underline{80} \\ 880 \end{array}$$

9. $800 \times 5 = 4{,}000$

10. $500 \times 3 = 1{,}500$

11. $700 \times 4 = 2{,}800$

12.
$$\begin{array}{r} {\scriptsize 11} \\ 138 \\ \underline{+274} \\ 412 \end{array}$$

13.
$$\begin{array}{r} {\scriptsize 11} \\ 705 \\ \underline{+398} \\ 1103 \end{array}$$

14.
$$\begin{array}{r} {\scriptsize 1} \\ 464 \\ \underline{+140} \\ 604 \end{array}$$

15. 40

16. 4,000

17. $11 \times 14 = 154$ sq ft

18. $2 \times 7 = 14$ days
$14 \times 12 = 168$ muffins

Lesson Test 24

1.
$$\begin{array}{r} 26 \\ \underline{\times 15} \\ 130 \\ \underline{26} \\ 390 \end{array}$$

2.
$$\begin{array}{r} 31 \\ \underline{\times 29} \\ 279 \\ \underline{62} \\ 899 \end{array}$$

3.
$$\begin{array}{r} 22 \\ \underline{\times 35} \\ 110 \\ \underline{66} \\ 770 \end{array}$$

4.
$$\begin{array}{r} 38 \\ \underline{\times 34} \\ {\scriptsize 3} \\ 122 \\ {\scriptsize 2} \\ \underline{94} \\ 1292 \end{array}$$

5.
$$\begin{array}{r} 28 \\ \underline{\times 39} \\ {\scriptsize 17} \\ 182 \\ {\scriptsize 2} \\ \underline{64} \\ 1092 \end{array}$$

6.
$$\begin{array}{r} 14 \\ \underline{\times 16} \\ {\scriptsize {}^{1}2} \\ 64 \\ \underline{14} \\ 224 \end{array}$$

7.
$$\begin{array}{r} 27 \\ \underline{\times 23} \\ {\scriptsize {}^{1}2} \\ 61 \\ {\scriptsize 1} \\ \underline{44} \\ 621 \end{array}$$

8.
$$\begin{array}{r} 16 \\ \underline{\times 24} \\ {\scriptstyle 2} \\ 44 \\ {\scriptstyle 1} \\ \underline{22} \\ 384 \end{array}$$

9. $100 \times 3 = 300$

10. $700 \times 8 = 5{,}600$

11. $200 \times 4 = 800$

12. 4, 8, 12, 16, 20, 24, 28, 32, 36, 40

13. $10 \times \underline{5} = 50$

14. $4 \times \underline{8} = 32$

15. $3 \times \underline{4} = 12$

16. $5 \times 48 = 240$ miles

17. $24 \times 36 = 864$ sq ft

18. $24 \times 31 = 744$ hours

Lesson Test 25

1.
$$\begin{array}{r} 312 \\ \underline{\times 53} \\ 936 \\ {\scriptstyle 1\ 1} \\ \underline{1550} \\ 16{,}536 \end{array}$$

2.
$$\begin{array}{r} 258 \\ \underline{\times 78} \\ {\scriptstyle 1} \\ {\scriptstyle 2\ 4\ 6} \\ 1604 \\ {\scriptstyle 1\ 3\ 5} \\ \underline{1456} \\ 20{,}124 \end{array}$$

3.
$$\begin{array}{r} 475 \\ \underline{\times 76} \\ {\scriptstyle 1\ 2\ 1} \\ {\scriptstyle 4\ 3} \\ 2420 \\ {\scriptstyle 4\ 3} \\ \underline{2895} \\ 36{,}100 \end{array}$$

4.
$$\begin{array}{r} 316 \\ \underline{\times \ 5} \\ {\scriptstyle 3} \\ \underline{1550} \\ 1{,}580 \end{array}$$

5.
$$\begin{array}{r} 34 \\ \underline{\times 96} \\ {\scriptstyle 1\ 1} \\ {\scriptstyle 2} \\ 184 \\ {\scriptstyle 3} \\ \underline{276} \\ 3{,}264 \end{array}$$

6.
$$\begin{array}{r} 81 \\ \underline{\times 11} \\ 81 \\ \underline{81} \\ 891 \end{array}$$

7. $12 \times 35 = 420$ sq ft

8. $35 + 12 + 35 + 12 = 94'$

9. $15 \times 18 = 270$ sq in

10. $15 + 18 + 15 + 18 = 66''$

11. $14 > 9$

12. $36 = 36$

13. $45 > 40$

14. 6, 12, 18, 24, 30, 36, 42, 48, 54, 60

15. $36 \times 108 = 3{,}888$ sq in

16. $52 \times \$250 = \$13{,}000$

17. $625 \times 55 = 34{,}375$ lb

18. $137 \times 31 = 4{,}247$ newspapers

Lesson Test 26

1. 1×6
2×3

2. 1×20
2×10
4×5

3. 1×18
2×9
3×6

4. 1×10
2×5

5. 1×15
3×5

6. 1×12
2×6
3×4

7. 1×9
3×3

8. 1×14
   2×7
9. $3 \times 25 = 75$
10. $14 \times 25 = 350$
11. $$\begin{array}{r} 185 \\ \times 13 \\ \hline {}^{1\,1} \\ {}^{2\,1} \\ 345 \\ 185 \\ \hline 2{,}405 \end{array}$$
12. $$\begin{array}{r} 693 \\ \times 21 \\ \hline {}^{1\,2} \\ 693 \\ {}^{1} \\ 1286 \\ \hline 14{,}553 \end{array}$$
13. $$\begin{array}{r} 564 \\ \times 48 \\ \hline {}^{1\,1} \\ {}^{4\,3} \\ 4082 \\ {}^{2\,1} \\ 2046 \\ \hline 27{,}072 \end{array}$$
14. $\frac{3}{6} = \frac{6}{12} = \frac{9}{18} = \frac{12}{24} = \frac{15}{30}$
15. $4 \times 6 = 24$ units
16. $23 \times 25 = 575$ cents
17. 1×16
    2×8
    4×4
18. $3 \times 25 = 75¢$
    $75¢ > 69¢$; yes

Lesson Test 27

1. six million, seven hundred one thousand, four hundred thirteen
2. five hundred seventy thousand, three hundred forty-eight
3. 402,519
4. 179,457,385
5. 204,213
6. $900{,}000{,}000 + 70{,}000{,}000 + 5{,}000{,}000 + 200{,}000 + 30{,}000 + 6{,}000 + 700 + 50 + 9$
7. $300{,}000 + 40{,}000 + 2{,}000 + 700 + 70 + 6$
8. $19 \times 16 = 304$ oz
9. $47 \times 16 = 752$ oz
10. $4 \times 16 = 64$ oz
11. 1×16
    2×8
    4×4
12. 1×20
    2×10
    4×5
13. 8×1
    2×4
14. 1×6
    2×3
15. $\frac{2}{9} = \frac{4}{18} = \frac{6}{27} = \frac{8}{36} = \frac{10}{45} = \frac{12}{54} = \frac{14}{63} = \frac{16}{72} = \frac{18}{81} = \frac{20}{90}$
16. $20 \times 16 = 320$ oz
17. $36 \times 16 = 576$ oz
18. $26 \times 16 = 416$ oz

Lesson Test 28

1. $$\begin{array}{r} 524 \\ \times 566 \\ \hline {}^{1\,2} \\ 3024 \\ {}^{1\,2} \\ 3024 \\ {}^{1\,2} \\ 2500 \\ \hline 296{,}584 \end{array}$$
2. $$\begin{array}{r} 8867 \\ \times \;\; 93 \\ \hline {}^{1\,1\,1\,1} \\ {}^{2\,1\,2} \\ 24481 \\ {}^{7\,5\,6} \\ 72243 \\ \hline 824.631 \end{array}$$

3.
```
       4461
      ×365
         1
       123
      20005
     123
     24466
     11
    12283
  1,628,265
```

4. 5,137,213
5. forty-four million,
   nine hundred thousand
6. 500,000,000 + 10,000,000 + 7,000,000 + 50,000 + 8,000 + 800
7. 1×12
   2×6
   3×4
8. 1×4
   2×2
9. 1×24
   2×12
   3×8
   4×6
10. 1×18
    2×9
    3×6
11. 6×8 = 48
12. 9×9 = 81
13. 7×7 = 49
14. 8×7 = 56
15. 125×125 = 15,625 sq ft
16. 1 widget per sec = 60 widgets per min
    60×60 = 3,600 widgets per hour
    3,600×8 = 28,800 widgets

Lesson Test 29

1. 1×6
   2×3; composite
2. 1×18
   2×9
   3×6; composite
3. 1×23; prime
4. 1×15
   3×5; composite
5. 1×19; prime
6. 1×12
   2×6
   3×4; composite
7. 10×12 = 120
8. 9×12 = 108
9. 6×12 = 72

10.
```
       485
      ×712
  1121
        11
       860
      485
    53
     2865
   345,320
```

11.
```
      5491
     ×  36
       1
       25
     30446
      12
    15273
   197,676
```

12.
```
       7123
       ×547
      1212
      49741
     1   1
     28482
      11
    35505
  3,896,281
```

13.
```
     8
    8 9 ¹2
   - 1 7 3
     7 1 9
```

14.
```
     11
     925
   +  86
   1,011
```

15.
```
      1
     204
   + 139
     343
```

16. 1×24
    2×12
    3×8
    4×6

17. $7 \times 12 = 84$ months

18. $6 \times 12 = 72$ eggs

Lesson Test 30

1. 5,280

2. 2,000

3. $3 \times 5{,}280 = 15{,}840$

4. $22 \times 2{,}000 = 44{,}000$

5. $7 \times 5{,}280 = 36{,}960$

6. $3 \times 2{,}000 = 6{,}000$

7. $12 \times 25 = 300$

8. $9 \times 16 = 144$

9. $11 \times 12 = 132$

10. $7 \times 12 = 84$

11.
$$\begin{array}{r} 392 \\ \underline{\times 234} \\ {}^{1} \\ {}^{1\,3} \\ 1268 \\ {}^{2} \\ 2976 \\ {}^{1} \\ \underline{684} \\ 91{,}728 \end{array}$$

12.
$$\begin{array}{r} 2638 \\ \underline{\times \quad 9} \\ {}^{1} \\ {}^{5\,2\,7} \\ \underline{{}^{1}18472} \\ 23{,}742 \end{array}$$

13.
$$\begin{array}{r} 4705 \\ \underline{\times \ 25} \\ {}^{3\ \ 2} \\ 20505 \\ {}^{1\ \ 1} \\ \underline{8400} \\ 117{,}625 \end{array}$$

14. $64 > 63$

15. $12 = 12$

16. $25 < 28$

17. $18 + 22 = 40$ tons

$40 \times 2{,}000 = 80{,}000$ lb

18. $6 \times 5{,}280 = 31{,}680$ ft

19. $5 \times 2{,}000 = 10{,}000$ lb

20. $14 \times 5{,}280 = 73{,}920$ ft

Unit Test IV

1.
$$\begin{array}{r} 21 \\ \underline{\times 48} \\ {}^{1} \\ 168 \\ \underline{84} \\ 1{,}008 \end{array}$$

2.
$$\begin{array}{r} 364 \\ \underline{\times 53} \\ {}^{1\,1} \\ 1982 \\ {}^{3\,2} \\ \underline{1500} \\ 19{,}292 \end{array}$$

3.
$$\begin{array}{r} 106 \\ \underline{\times 789} \\ {}^{1\,5} \\ 904 \\ {}^{1\,4} \\ 808 \\ {}^{1\,4} \\ \underline{702} \\ 83{,}634 \end{array}$$

4.
$$\begin{array}{r} 1357 \\ \underline{\times \quad 6} \\ {}^{1} \\ {}^{1\,3\,4} \\ \underline{6802} \\ 8{,}142 \end{array}$$

5.
$$\begin{array}{r} 2843 \\ \underline{\times \ 75} \\ {}^{1\,1\,1} \\ {}^{4\,2\,1} \\ 10005 \\ {}^{5\,2\,2} \\ \underline{14681} \\ 213{,}225 \end{array}$$

6.
$$\begin{array}{r} 4561 \\ \underline{\times \ 32} \\ {}^{1\,1} \\ {}^{1}8022 \\ {}^{1\,1} \\ \underline{12583} \\ 145{,}952 \end{array}$$

7. 1×9

3×3; composite

8. 1×12

2×6

3×4; composite

9. 1×7; prime
10. $8 \times 25 = 200$
11. $10 \times 16 = 160$
12. $3 \times 12 = 36$
13. $6 \times 5{,}280 = 31{,}680$
14. $2 \times 2{,}000 = 4{,}000$
15. $25 \times 12 = 300$
16. 90
17. 100
18. 5,000
19. 4,568
20. $2{,}000{,}000 + 300{,}000 + 90{,}000 + 1{,}000 + 600$

Final Test

1.
$$\begin{array}{r} 85 \\ \underline{\times 26} \\ {\scriptstyle 13} \\ 480 \\ {\scriptstyle 11} \\ \underline{160} \\ 2{,}210 \end{array}$$

2.
$$\begin{array}{r} 421 \\ \underline{\times 73} \\ {\scriptstyle 1} \\ 1263 \\ {\scriptstyle 11} \\ \underline{2847} \\ 30{,}733 \end{array}$$

3.
$$\begin{array}{r} 509 \\ \underline{\times 636} \\ {\scriptstyle 15} \\ 3004 \\ {\scriptstyle 1\ \ 2} \\ 1507 \\ {\scriptstyle 5} \\ \underline{3004} \\ 323{,}724 \end{array}$$

4.
$$\begin{array}{r} 7546 \\ \underline{\times \quad 8} \\ {\scriptstyle 434} \\ \underline{{}^{1}56028} \\ 60{,}368 \end{array}$$

5.
$$\begin{array}{r} 3482 \\ \underline{\times \quad 59} \\ {\scriptstyle 111} \\ {\scriptstyle 371} \\ 27628 \\ {\scriptstyle 241} \\ \underline{15000} \\ 205{,}438 \end{array}$$

6.
$$\begin{array}{r} 6187 \\ \underline{\times 467} \\ {\scriptstyle 1131} \\ {\scriptstyle 54} \\ 42769 \\ {\scriptstyle 44} \\ 36682 \\ {\scriptstyle 32} \\ \underline{24428} \\ 2{,}889{,}329 \end{array}$$

7. $31 \times 72 = 2{,}232$ sq ft
8. $31 + 72 + 31 + 72 = 206'$
9. $8 \times \underline{8} = 64$
10. $9 \times \underline{7} = 63$
11. $10 \times \underline{10} = 100$
12. 1×16
    2×8
    4×4; composite
13. 1×7; prime
14. 1×9
    3×3; composite
15. $12 = 12$
16. $72 > 60$
17. $42 < 45$

18.
$$\begin{array}{r} {\scriptstyle 1} \\ 92 \\ 21 \\ 48 \\ \underline{+17} \\ 178 \end{array}$$

19.
$$\begin{array}{r} {\scriptstyle 1} \\ 163 \\ \underline{+54} \\ 217 \end{array}$$

20.
$$\begin{array}{r} 815 \\ \underline{+482} \\ 1{,}297 \end{array}$$

21.
```
   5
  3 6̸ ¹0
-   3 7
  3 2 3
```

22.
```
  4
  5̸ ¹29
- 168
  3 6 1
```

23.
```
  3 9
  4̸ ¹0̸ ¹2
- 2 9 3
  1 0 9
```

24. $6 \times 2 = 12$
25. $8 \times 10 = 80$
26. $9 \times 3 = 27$
27. $5 \times 3 = 15$
28. $10 \times 5 = 50$
29. $7 \times 4 = 28$
30. $2 \times 4 = 8$
31. $4 \times 8 = 32$
32. $3 \times 16 = 48$
33. $6 \times 25 = 150$
34. $2 \times 5{,}280 = 10{,}560$
35. $1 \times 2{,}000 = 2{,}000$
36. $20 \times 40 = 800$ sq ft
37. $500 \times 3 = 1{,}500$ mi
38. 3,000
39. 1,271,028
40. $5{,}000{,}000 + 600{,}000 + 80{,}000 + 1{,}000 + 900$

Symbols and Tables

SYMBOLS

Symbol	Meaning
=	equals
+	plus
–	minus
×	times/multiply
·	times
( )( )	times
¢	cents
$	dollars
'	foot
"	inch
<	less than
>	greater than

TIME

60 seconds = 1 minute
60 minutes = 1 hour
1 week = 7 days
1 year = 365 days (366 in a leap year)
1 year = 52 weeks
1 year = 12 months
1 decade = 10 years
1 century = 100 years

MONEY

1 penny = 1 cent (1¢)
1 nickel = 5 cents (5¢)
1 dime = 10 cents (10¢)
1 quarter = 25 cents (25¢)
1 dollar = 100 cents (100¢ or $1.00)
1 dollar = 4 quarters

LABELS FOR PARTS OF PROBLEMS

Addition

25	addend
+16	addend
41	sum

Multiplication

3 3	multiplicand (factor)
× 5	multiplier (factor)
1 6 5	product

Subtraction

4 5	minuend
– 2 2	subtrahend
2 3	difference

MEASUREMENT

1 quart (qt) = 2 pints (pt)
1 gallon (gal) = 8 pints (pt)
1 gallon (gal) = 4 quarts (qt)
1 tablespoon (Tbsp) = 3 teaspoons (tsp)
1 foot (ft) = 12 inches (in)
1 yard (yd) = 3 feet (ft)
1 mile (mi) = 5,280 feet (ft)
1 pound (lb) = 16 ounces (oz)
1 ton = 2,000 pounds (lb)
1 dozen = 12

PLACE-VALUE NOTATION

931,452 = 900,000 + 30,000 +1,000 + 400 + 50 + 2

Glossary

A-C

area - the measure of the space covered by a plane shape, expressed in square units

Associative Property - a property that states that the way terms are grouped in an addition expression does not affect the result

century - one hundred years

Commutative Property - a property that states that the order in which numbers are added does not affect the result

composite number - a number with more than two factors

D-E

decade - ten years

denominator - the bottom number in a fraction, which shows the number of parts in the whole

dimension - a measurement in a particular direction (length, width, height, depth)

equation - a mathematical statement that uses an equal sign to show that two expressions have the same value

estimate - a close approximation of an actual value

even number - any number that can be evenly divided by two

F-I

factor - (n) a whole number that multiplies with another to form a product; (v) to find the factors of a given product

fraction - a number indicating part of a whole

hexagon - a polygon with six sides

inequality - a mathematical statement showing that two expressions have different values

J-O

multiplicand - in multiplication, the factor that is being repeated

multiplier - in multiplication, the number that indicates how many times the other factor is being repeated

numerator - the top number in a fraction, which shows the number of parts being considered

octagon - a polygon with eight sides

odd number - any number that cannot be evenly divided by two

P-R

pentagon - a polygon with five sides

perimeter - the distance around a polygon

place value - the position of a digit which indicates its assigned value

place-value notation - a way of writing numbers that shows the place value of each digit

prime number - a number that has only two factors: one and itself

product - the result when numbers are multiplied

quadrilateral - a polygon with four sides

rectangle - a quadrilateral with two pairs of opposite parallel sides and four right angles

regrouping - composing or decomposing groups of ten when adding or subtracting

right angle - an angle measuring 90 degrees

rounding - replacing a number with another that has approximately the same value but is easier to use

S-Z

skip counting - counting forward or backward by multiples of a number other than one

square - a quadrilateral in which the four sides are perpendicular and congruent

triangle - a polygon with three straight sides

unit - the place in a place-value system representing numbers less than the base

unknown - a specific quantity that has not yet been determined, usually represented by a letter

Master Index for General Math

This index lists the levels at which main topics are presented in the instruction manuals for *Primer* through *Zeta.* For more detail, see the description of each level at mathusee.com. (Many of these topics are also reviewed in subsequent student books.)

Gamma Index